HOSPITALS

医院建筑

于 萍 编 常文心 译

辽宁科学技术出版社

My Design Experience

我的设计经验与体会

I have been devoting my life to medical architecture over 35 years.

This "special art" must be incorporated with "human life" to create works which combine human scale and functions. Medical architecture is an "organism". As history accumulates, society changes and medical science develops, medical architecture is growing up continually and reflecting different era's requirements. Working deeply in this area, you have to be extremely observant and reflective, as well as maintain a selfless and passionate soul.

I like to get involved with sky, earth and human beings; therefore, astronomy, Greek mythology and mountain-climbing have always been my inspirations. They broaden my view, accumulate my energy of self-breaking and enhance my architectural judgment and ability.

In the process of working on medical architecture, "communication" plays an important role, especially when denling with hospital staff with selfish departmentalism. Once, an obstetrician required the delivery room to be large and separated by curtains, which would be convenient for delivery. However, my experience and

医疗建筑是我一生的志业，至今已走过35个寒暑。

这门“特别的艺术”，必须深刻结合“人间生活”，才能创造出兼具人性和功能的作品。医疗建筑是一“有机体”，伴随着历史累积、社会变迁与医疗科技的进步，不断地成长，并反馈出不同的时代需求，深入这个领域，需要极端冷静的观察、思考和一颗无私、激情的灵魂。

我喜欢感受天、地、人的真实，因此天文、希腊神话、登山一直是我的灵感泉源，它使我的视野更加辽阔，自我突破的能量愈加积累，提升了建筑师的判断力和专业能力。

医疗建筑的过程“沟通”非常重要，特别当对象是部门本位主义浓厚的医院人员。曾有一次，妇产科医师要求产房设计要大且宽敞，用帘子分隔，方便接生。经验与判断力告诉我，纵是站在使用者与设计者的对话天秤上，这是绝对不可妥协的！因为产妇在承受巨大的生产痛苦时，耳边还充塞其他产妇的呻吟哀号，其内心的恐惧感可想而知。

医疗建筑的演变与大环境密切相关。多年来，我从事医疗建筑设计，有自我要求的理念基础：首先要有“弹性”，空间须默认弹性以符合病患数目的增长，土地资源有限，必须妥善合理之运用。其次，是“竞争优势”，医疗工艺、动线合理、相关部门专科资源共享等，病人才不致如入迷宫，加之绿色节能等规划，医院本身才有竞争优势。第三是“人性化”，并与创新作结合，从实务应用出发思考，让病人在医院时空中，感到亲切、放松与信任，例如公共区域的人文空间，关照精神层面的需求，增加人性化服务。第四是“重视防范院内感染”，智能配置空间，利用正负压差控制气流，清浊空气分流。唯有掌握人、阳光、空气、水与环境疗愈、物品动线、分明清污才能收最佳感染防制之效。

在所有建筑设计类型中，“专业化”这项特点，点出了医疗建筑与其他建筑之最大差异，我从事医疗建筑设计，深觉命运蕴涵其中，意义非凡，努力的脚步，贯彻始终。

judgment told me that it is non-negotiable, on matter from the point of user or designer. When a parturient is experiencing enormous labour pain, you can imagine she would be more panic hearing other parturients' crying.

The evolvement of medical architecture is closely connected with environment. For years of my practice in medical architecture design, I have my demanding concept. First, space must be equipped with flexibility to match the increasing number of patients, as the land resource is limited and we have to use it properly. Second, the hospital has to contain "competitive edge", such as medical technology, reasonable circulation and professional resource sharing, in order to guide the patients easily. In addition of green features and energy strategy, the hospital will be competitive. Third, the hospital need to be humane. Combining human scales and innovations and starting from practical application, the hospital will make patients feel intimacy, relaxation and confidence. For example, the hospital should take care of cultural space of the public area, pay attention to patients' spiritual needs and add humanised services. Forth, the hospital must avoid internal infection. The design should configure spaces intelligently and use pressure difference to control air current and to separate the clean and dirty air. Only when human, sunlight, air, water, environment, materials circulation and separation are all mastered in hand, the best infection avoidance can be achieved.

In all architecture types, "professional" is the feature which differentiates medical architecture and other types of architecture. In my experience of working on medical architecture, I have found my fortune and its outstanding meanings. My studious steps never stop.

Hsu Chang-Ch
许常吉 建筑师

Contents
目 录

General Hospital 综合医院

Rehabilitation Hospital 疗养医院

Children's Hospital 儿童医院

阿克什胡斯大学医院 Akershus University Hospital

Location: Oslo, Norway **Completion date:** 2008 **Designer:** C. F. Møller Architects **Photographer:** Torben Eskerod **Area:** 137,000 m^2

项目地点：挪威，奥斯陆 完成时间：2008年 设计师：C.F.穆勒建筑事务所 摄影师：托本·艾斯克罗德 面积：137,000平方米

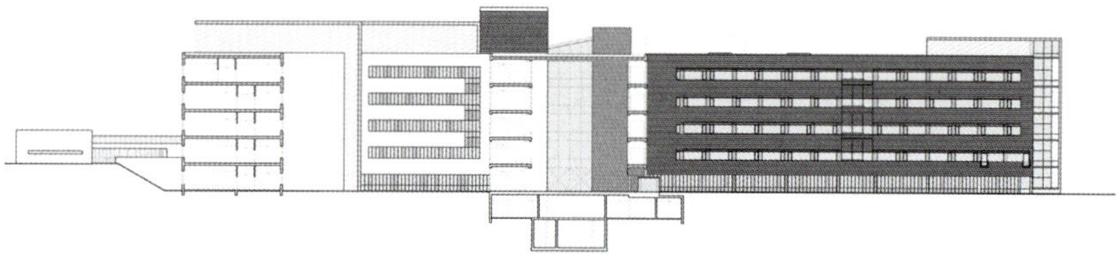

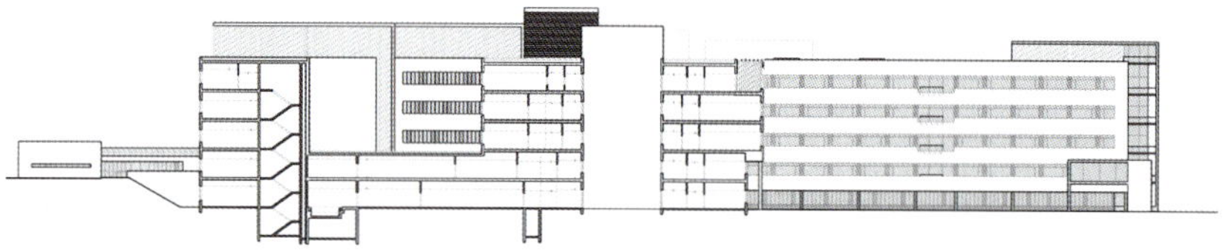

The new university hospital is not a traditional institutional construction; it is a friendly, informal place with open, well-structured surroundings which present a welcoming aspect to patients and their families. Akershus University Hospital has been designed to emphasise security and clarity in experientially rich surroundings, where everyday functions and well-known materials are integrated into the hospital's structure.

Although the individual parts of the development each have their own material expression and the material expression of the development varies, nonetheless it is united into a whole by means of a general architectural theme centred on panels and transparency. In this way, a unity is created between the individual parts of the complex, which thereby receive a subtle effect of transparency and depth.

A glass-roofed main thoroughfare, in which wood is the dominating material, links the various buildings and departments. The "glass street" begins in the welcoming foyer of the arrivals area, where the main reception desk receives visitors, and concludes in the foyer and separate arrivals area of the children's department.

In the glass street, the central element in the development, the various materials are united in an overall composition, in which the large coloured panels designed by the Icelandic artist Birgir Andrésson form a natural element and provide a "palette" for the colour scheme of the hospital.

The glass street has a town-like structure, with public and semi-public zones defined as squares and open spaces, offering the everyday functions of a town: church, pharmacy, hairdresser, florist, café and kiosk, as well as traffic nodes and other services for the benefit of patients, relatives and staff. In natural continuation of these functions, a number of other services, such as health information, polyclinics and out-patient surgeries, are located near the street level of the thoroughfare.

The hospital's structure helps to ensure that the patient remains the natural focus in the physical design, despite the strict and demanding logistical requirements which underlie all hospital constructions. Just as the overall complex is made up of clear and comprehensible units, so the individual wards are built up from smaller elements. The wards are centred around four so-called courtyards which ensure a well-defined daily life for the patients, with a manageable level of social contact, assisted by a clear staff interface.

The wards of the children's department are equipped with windows which give the children and young people individual views of both the sky and the surrounding greenery from their beds. The well-equipped facilities for parents secure excellent contact between the children and their families.

新建的大学医院并不是一座传统意义上的医疗设施，它亲切友好的环境和开放合理的周边让患者和家属们感受
尽管建筑的各部分都采用了不同的材料，各具特色，但是它们通过综合的建筑主题——面板和透明度——形成了统一的主体。这样一来，建筑的各个部分都拥有了精妙的透明度和深度效果。
以木材为主要材料的玻璃屋顶大道连接了各个楼和部门。“玻璃街”起于大厅的到达区，通往大厅内部和儿科的各个到达区。
在玻璃街——项目的核心元素里，各种材料得到了统一，巨大的彩色玻璃板（由冰岛艺术家波尔吉尔·安德森设计）显得十分自然，形成了医院色彩主题的调色板。
玻璃街有着城镇的结构，公共和半公共区域通过广场和开放空间来划分，也提供了城镇的基本设施：教堂、

1. The geometrical structure looks clean and lively
2. Main entrance of the hospital
3. The grey inpatient building
4. The façade uses structural glass

1. 几何形状建筑结构设计简洁明快
2. 医院入口处
3. 灰色的住院部大楼
4. 建筑外观采用了大块玻璃材料

1

2

3

4

1

2

3

药房、理发店、花店、咖啡厅和电话亭，还拥有一些交通节点和病患、家属、员工服务设施。此外，一系列的其他服务设施，如健康咨询、综合门诊和门诊手术等服务都靠近街面。

医院的结构保证了患者与建筑的物理设计之间的联系，也考虑了医院建筑要求严格的后勤需求。就像整体建筑由清晰明确的部分组成一样，个人病房也由小型元素组成。病房围绕着四个庭院展开，为患者提供优质的日常服务和正常的社交环境。

儿科病房的窗户让儿童和年轻人能在病床上看到天空和周边绿地的风景。为家长所准备的设施保证了儿童和家庭的良好联系。

Award:

Prize in Restricted Architectural Competition 2000

Building Better Healthcare Award 2009 Best International Design

获奖情况:

2000年内部建筑竞赛奖

2009年“打造更好的医疗设施”奖，国际最佳奖

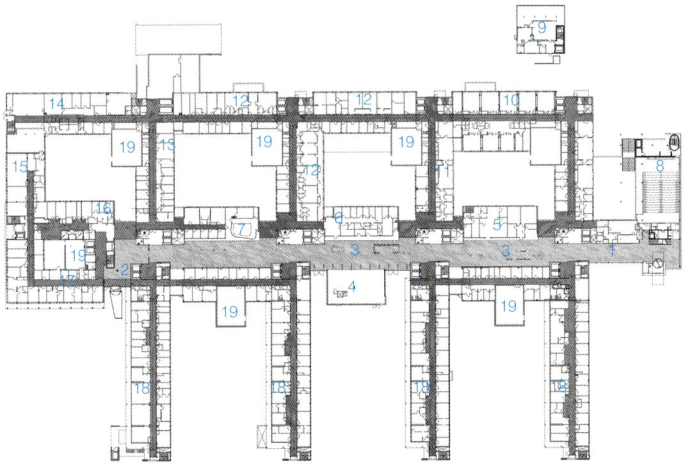

1. Main entrance and reception
2. Children's entrance and reception
3. Glass covered mall
4. Café/kiosk
5. Pharmacy
6. Blood sampling
7. Church
8. Auditorium
9. Chapel
10. Day surgery
11. Anaesthesia/recovery
12. Diagnostic imaging
13. Pain clinic
14. Physiotherapy/rehabilitation
15. Rehabilitation children
16. Sampling children
17. Polyclinic children
18. Out-patient departments polyclinics
19. Technical towers

1. 主入口和接待处
2. 儿科入口和接待处
3. 玻璃罩林荫路
4. 咖啡亭/报亭
5. 药房
6. 取血处
7. 教堂
8. 礼堂
9. 礼拜堂
10. 日间手术室
11. 麻醉/恢复室
12. 影像诊断室
13. 疼痛门诊
14. 理疗/康复室
15. 儿科康复室
16. 儿科取血处
17. 儿科综合门诊
18. 综合门诊部
19. 技术室

1. The side view of the hospital
2. The outer space is a green area for the patients
3. A night view of the building

1. 医院侧面
2. 住院部外围为病患提供了大块绿地
3. 建筑夜景

2

3

4

1. The interior applied numerous wood structures
2. The glass roof connects different departments inside the building
3. The selection of wood creates a warm and natural sense
4. The interior passage

1. 医院内部采用大量木制结构
2. 玻璃屋顶连接了大楼内部各个部门
3. 木材的使用带来了自然亲切感
4. 内部通道

Sengebygg 3
5. etasje
4. etasje
3. etasje
2. etasje
1. etasje
0199
20
ringla heimsins

1

2

3

4

1. Elevator lobby, with robotic automated AGV unit transporting supplies
2. The interior public sitting area
3. The underground supply system includes automated AGV units which handle transports and logistics within the hospital
4. Detail from the underground supplies facility

1. 电梯大堂，自动无人搬运车负责运输供给品
2. 建筑内部的公共休息区
3. 地下供应系统包括自动无人搬运车，负责医院内部的运输和物流
4. 地下供应设施细部图

克莱蒙费朗医院 Clermont-Ferrand Hospital

Location: Clermont-Ferrand, France **Completion date:** 2009 **Designer:** Groupe-6 **Photographer:** Luc Boegly **Area:** 70,000 m^2

项目地点：法国，克莱蒙费朗 完成时间：2009年 设计师：6队设计公司 摄影师：卢克・博尔格利 面积：70,000平方米

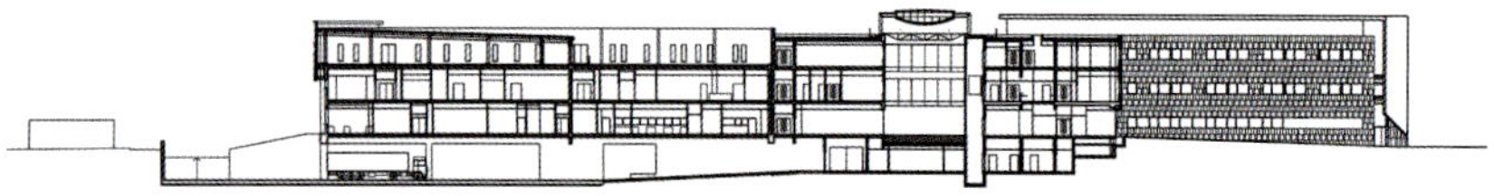

With a capacity of 565 beds over 70,000 square metres, more than a healthcare project the Clermont-Ferrand Hospital (CHUE) is an urban project which offers opportunities for the extension of the city beyond today's frontiers. Replacing the old "Hotel-Dieu" erected in the 18th century in the town centre, the new hospital is built on the former site of the Michelin factory. Located at the boundaries of the city, in a quarter undergoing radical transformation, the CHUE acts as a proper "urban generator", and becomes the anchor of a new town designed to become a future healthcare centre. With its monumental entry, like a metallic sail floating over the square, the building acts as a landmark and meeting point for the new district. Genuine technical prowess, it is made of five pillars supporting the 1,500 square metres of the metallic cover. With its planted nave, it extends the existing urban frame while defining new axes of circulation. Everything in the CHUE borrows from the urban typology and vocabulary: places, squares, streets, alleyways... the hospital is already in the city. The internal green passage of 250 metres long is punctuated by reading and resting zones, treated like a semi-public, pedestrian and commercial space; in other words: an urban street.
Its lower ground plus two wards levels, coupled with its fragmented volumetry, complete this primary "macro" structure and offer more intimist and domestic spaces. The scales management and the spaces hierarchy show a functional organisation based on a pole system and a stratification of activities.
With a lifetime of 40 years, healthcare facilities must be flexible and competitive. Here the architects have deployed a real ambition, with an efficient and fast construction site, a nave working like a ventilating buffer space, and a structure allowing potentially a 20% extension.

克莱蒙费朗医院共有565个床位，总面积7万多平方米，不仅是一家医疗机构，更是一个城市扩建工程的重要城市规划项目。
新医院替代了市中心建于18世纪的“天主酒店”，建在米其林工厂的原址上。医院所在的城市边缘正经历着巨大的变化，作为一个合适的“城市生成器”，医院成为了新城的标志性建筑。宏伟的入口仿佛广场上的一面金属船帆，医院是新区的地标和集会场所。在新技术的支持下，5根柱子支撑着1,500平方米的金属外壳。绿化中庭延伸了原有的城市框架，界定了新的流通路径中轴。医院的设计借鉴了城市规划，包括广场、街道、小巷等，整个医院就是一座城市。250米长的室内绿色通道里点缀着阅读和休息区，形成了半公共的步行商业空间，或者说，一条城市街道。
一楼和上面的两层病房与独立的体量共同组成了医院的基本结构，提供了私密而舒适的空间。规模经营和空间谱系体现了以极点系统和活动层理为基础的功能设置。
医疗设施的使用年限为40年，必须灵活而耐用。建筑师运用高效率而快速的工程方法，通过中庭提供了一个通风缓冲区，整体可进行20%的扩建。

1. A side view
2. A metallic sail sits outside the monumental entry
3. The metallic roof is supported by five pillars

1. 医院侧立面
2. 入口外侧的金属船帆气势恢宏
3. 五根柱子支撑着金属顶棚

1

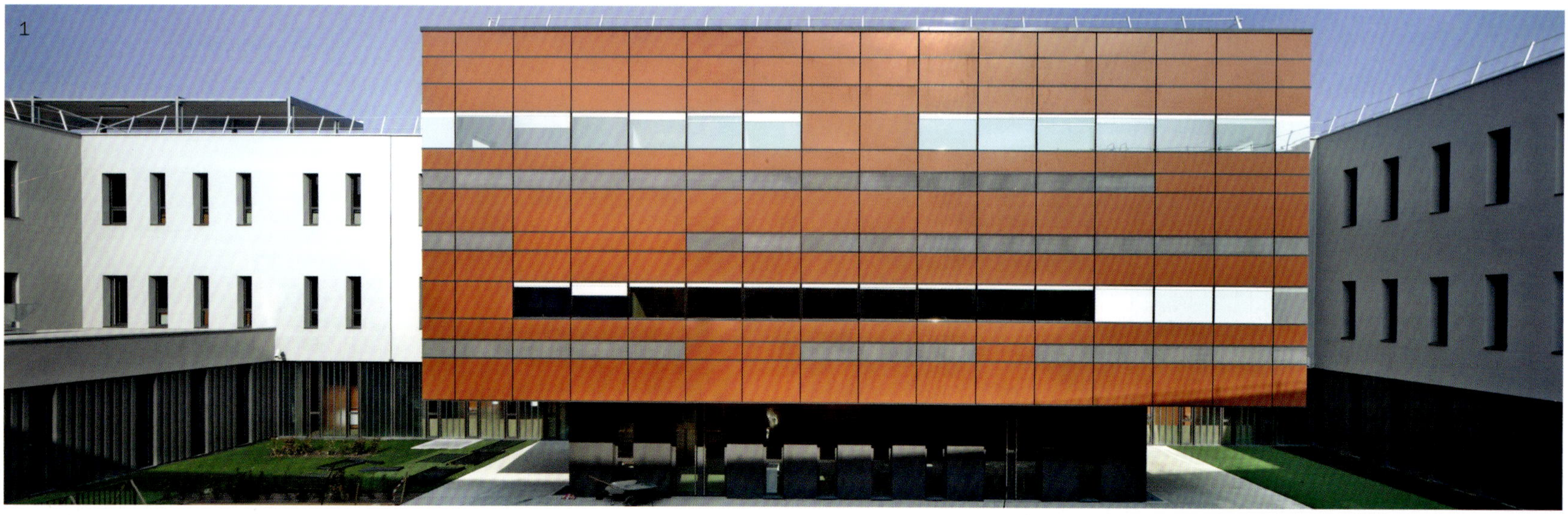

2

3

1

2

3

4

1. There is an exterior stair on one side of the building
2. Matching with glass, the red wall enhanced the building's height
3. The French windows allow in abundant daylight
4. The interior of the hall combined well with the façade

1. 建筑侧面附有外部楼梯通道
2. 红色外墙搭配玻璃材料提升了建筑的亮度
3. 高大的落地玻璃使大厅内部采光良好
4. 大厅内部与外墙嵌入融合完美

1. Main entrance
2. Lobby
3. Reception
4. Sickrooms
5. Consulting rooms
6. Recovery
7. Nurse stations
8. Lift lobby
9. Waiting areas
10. Staff rooms
11. Inter courtyard

1. 主入口
2. 大堂
3. 接待处
4. 病房
5. 诊疗室
6. 恢复室
7. 护理站
8. 电梯大堂
9. 等候区
10. 员工休息室
11. 室内庭院

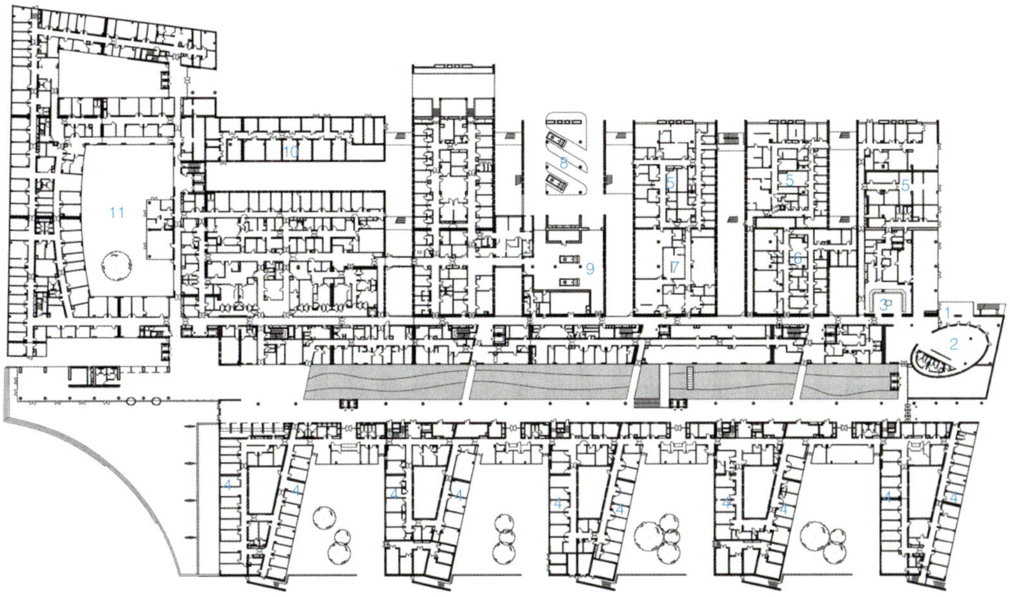

1. It is open and spacious inside
2. The elegant interior landscape
3. The central hall is peaceful and quiet
4. The interior design is close to nature

1. 医院内部开阔宽敞
2. 优雅的室内景观环境
3. 中央大厅宁静平和
4. 室内设计贴近自然

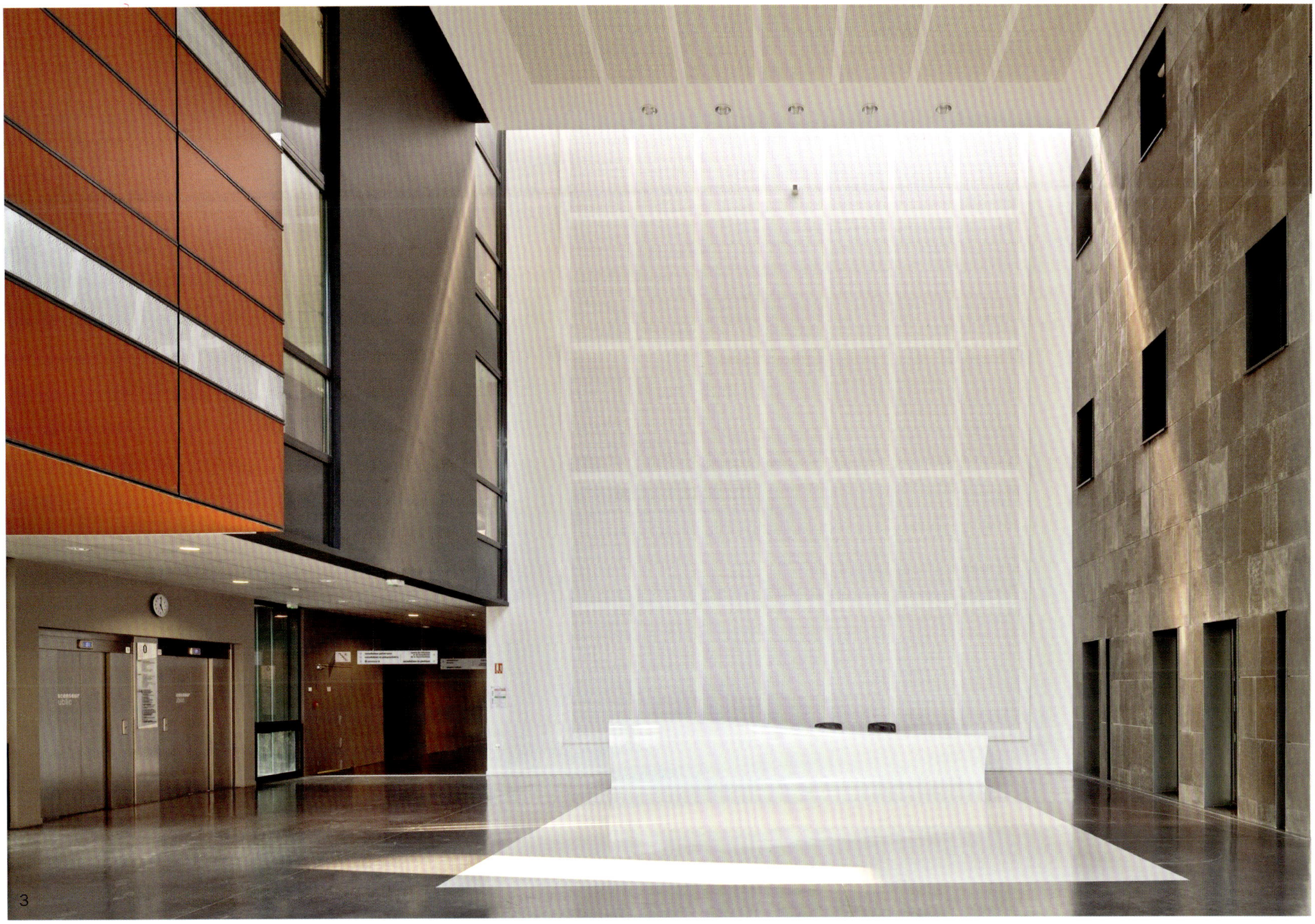

哥伦比亚亚洲医疗中心 Columbia Asia Medical Centre

Location: Johor, Malaysia **Completion date:** 2010 **Designer:** Environmental Design Practice Sdn. Bhd. **Photographer:** RGB Studio Photography, Kuala Lumpur **Area:** 10,200 m^2

项目地点：马来西亚，柔佛 完成时间：2010年 设计师：环境设计实践公司 摄影师：RGB摄影工作室，吉隆坡 面积：10,200平方米

The brief asked for a hospital that does not look like a typical hospital. It should exude confidence in users as a place of science and precision in healthcare facilities. The Columbia Asia Medical Centre is located in Nusajaya Medical Park which is part of the new special economic zone of Malaysia located at the southern-most tip of Peninsular Malaysia. It is to be one of the first buildings to be developed in the soon-to-be major regional commercial hub equipped with "high-tech" first-world facilities.

Unlike most of the conventional local hospitals that try to create a vernacular or tropical architectural language, this hospital makes no attempt to follow the old paradigm hospital design in the region. It uses curtain wall to maximise natural daylight. Visually pleasing in proportion, the cladding used was a bright silver composite aluminium panels and blue tinted glass to give a modern curtain wall look with a clean sleek image. This functional high-tech appearance aims to impart a sense of confidence in the healing technology. Since the structural element employed was relatively simple, the architect derived elegance of the building from the intricate language of detailing work. The approach was to create a balance between satisfying the complex functional requirements of an efficient compact hospital and the creation of an easily recognisable and "iconic" architecture for the owner as well as addressing the surrounding context of the medical park.

The interior design of the hospital, on the other hand, takes on a softer approach whilst maintaining a "homely" feel. Again, the aesthetics are achieved by the appropriate use of intricate detailing in the joinery work. Natural lighting is therapeutic and here it is abundant in the wards as well as the common areas in the building.

The prototype design took the shape of an efficient rectangular shape, where generally the First floor is for the patient care. All the patient rooms are placed here along the perimeter with external windows for filtered daylight, and possible natural ventilation in case of power failure. Ground floor is dedicated to diagnostic, treatment and administrative functions. All supporting services and back-of-house functions occupy the only basement level. All the mechanical plants and water tanks are planned to centralise entirely on the covered roof level, which forms the mass necessarily for the required façade proportion consideration. This enclosed roof level becomes an interstitial space insulating the habitable floor below, as such reducing vertical heat load, which is predominant in tropical climate. By so doing, all habitable floors are also free of mechanical rooms, therefore reducing heat, noise and contamination risk.

设计要求打造一个不同寻常的医院，它应该散发自信，体现医疗机构的科技性和精确度。哥伦比亚亚洲医疗中心位于努沙再也医疗园，是马来西亚新经济特区的一部分，地处马来西亚半岛的南端。医院是这个未来区域商业中心第一座建筑，中心将配备世界一流的高科技设施。

大多数传统的本地医院试图打造本地或者热带风情的建筑，这座医院则颠覆了该区域的传统模式。它使用玻璃幕墙来最大化自然采光。明亮的银色复合铝板和蓝色玻璃让幕墙显得现代而简洁、干练。这种高科技感增添了人们对医院医疗技术的信心。由于建筑的整体结构相对简单，建筑师把精致感放在了细节上面。他们试图在简练而功能齐全的医疗建筑和为园区打造一座地标性建筑之间达成一个平衡。

另一方面，医院的室内设计相对柔和，给人以家的感觉。美观度同样体现在精致的木工细节上。由于自然采光有益于健康，病房和公共区域的采光都十分充足。

建筑采用了高性能的长方体造型，二楼是病患护理区。所有的病房都沿着窗口设置，保证了自然采光和通风。一楼是诊断、治疗和行政区。所有的后勤服务和辅助设施都设在地下室。机械设备和水草全部设在封闭式屋顶，使建筑的比例十分得当。封闭式屋顶相当于楼下的缓冲层，减少了热带区域常见的垂直热量流失。这样一来，地下的楼层也不需要机械室，减少了热量、噪声和污染。

1. A night view of the medical centre
2. The colourful main building
3. The façade consists of silver aluminium sheet and blue glass
4. The design is modern and clean

1. 医疗中心夜景
2. 色彩丰富的建筑主体
3. 银色铝板和蓝色玻璃构成了建筑外墙
4. 建筑设计现代简洁

3

4

1

1. It is warm and bright inside
2. The interior space has a good supply of daylight

1. 内部设计温馨明快
2. 室内采光良好

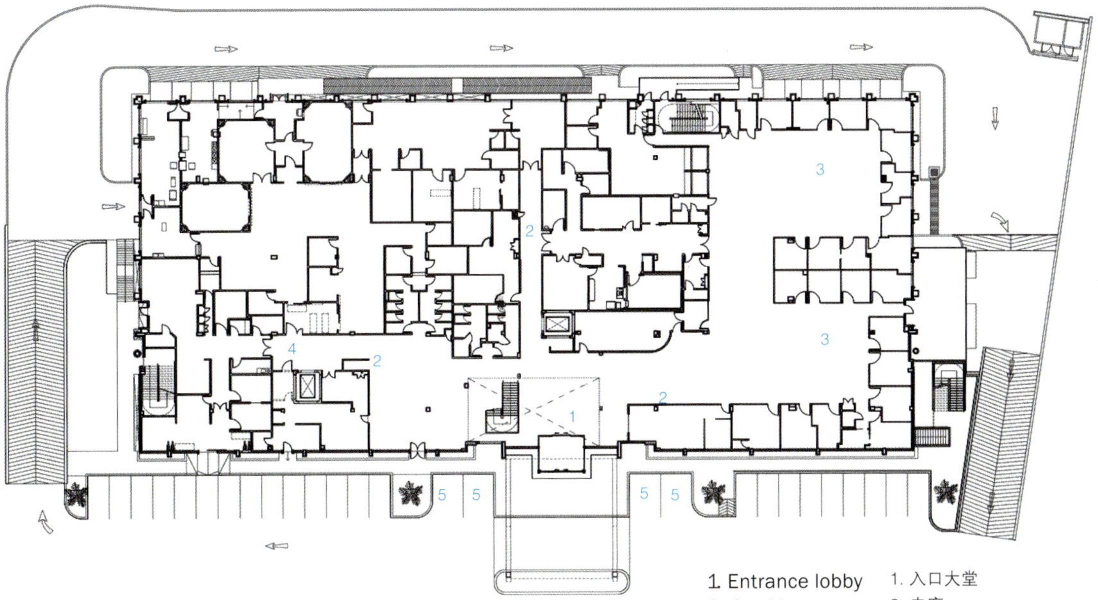

1. Entrance lobby
2. Corridor
3. Waiting area
4. Lift lobby
5. Disable parking

1. 入口大堂
2. 走廊
3. 等候区
4. 电梯大堂
5. 残障停车场

1

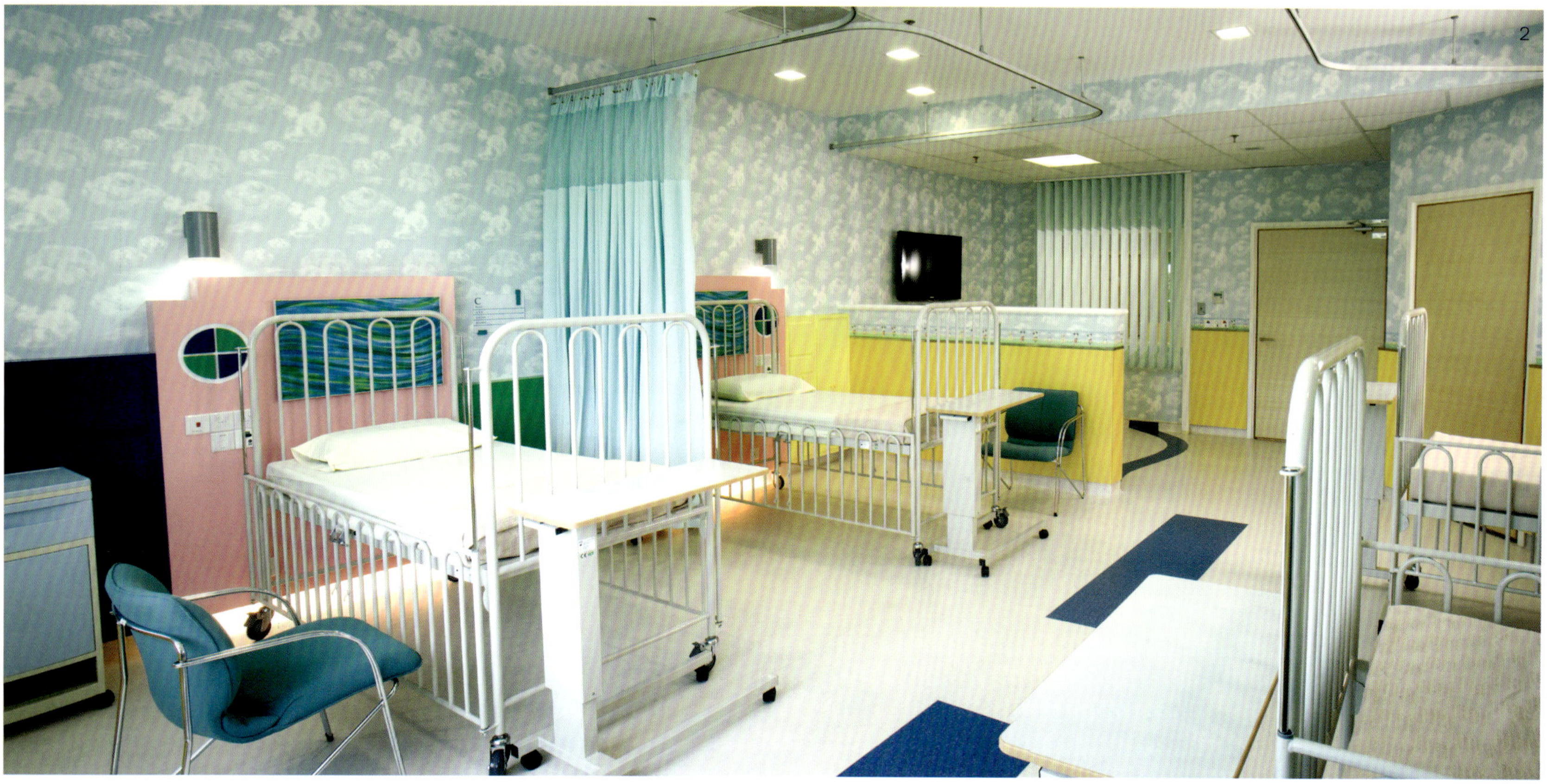
2

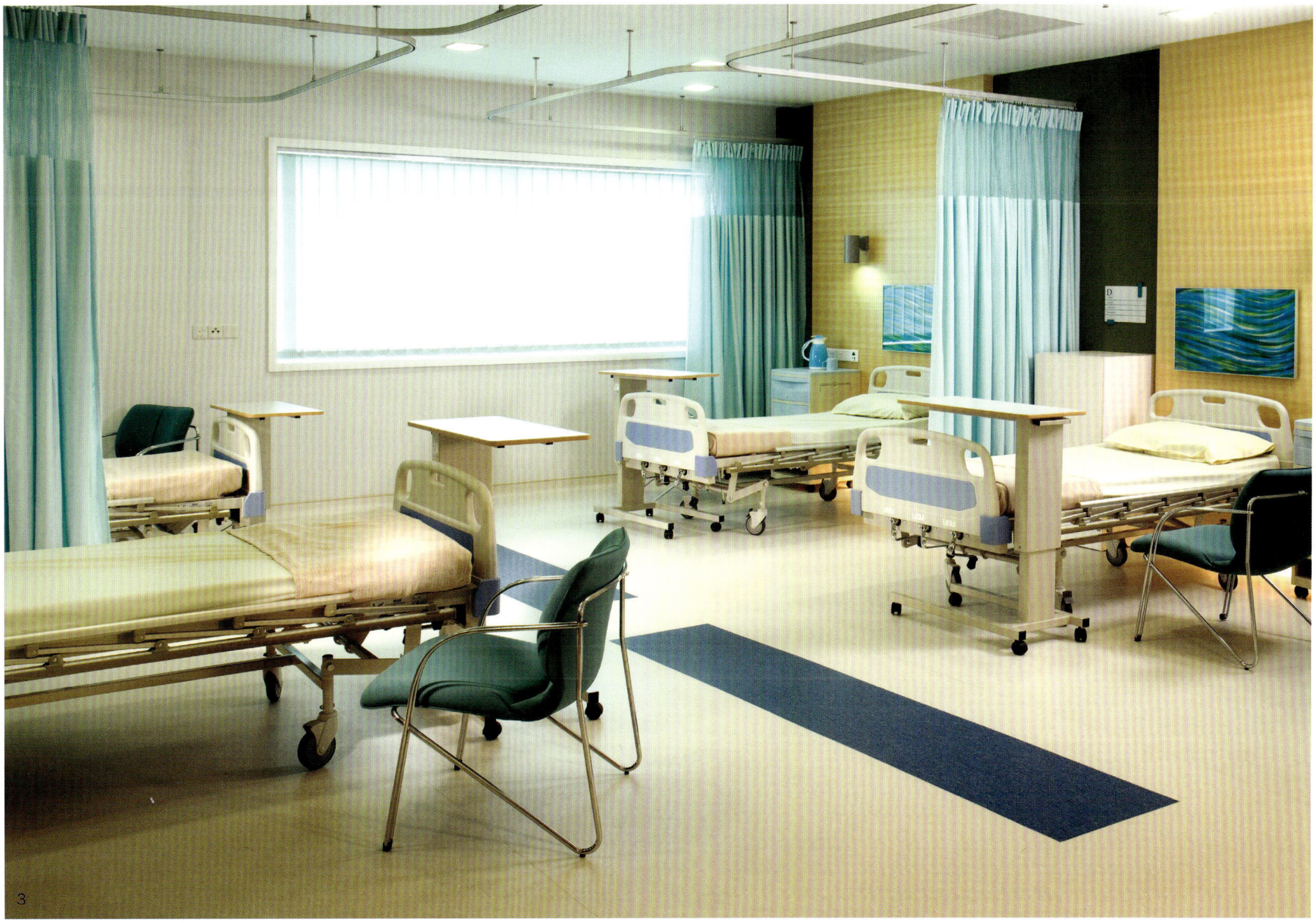

1. The open and clean patient room
2. The warm and comfortable children room
3. The interior design emphasises daylight and natural ventilation

1. 宽敞整洁的病房
2. 温馨的儿童病房
3. 室内设计注重自然采光和通风

北方社区医院扩建工程 Community Hospital North Expansion

Location: Indianapolis, Indiana, USA **Completion date:** 2007 **Designer:** RTKL **Photographer:** Jeffrey Totaro

项目地点：美国，印第安纳波利斯 完成时间：2007年 设计师：RTKL事务所 摄影师：杰弗瑞·托拉多

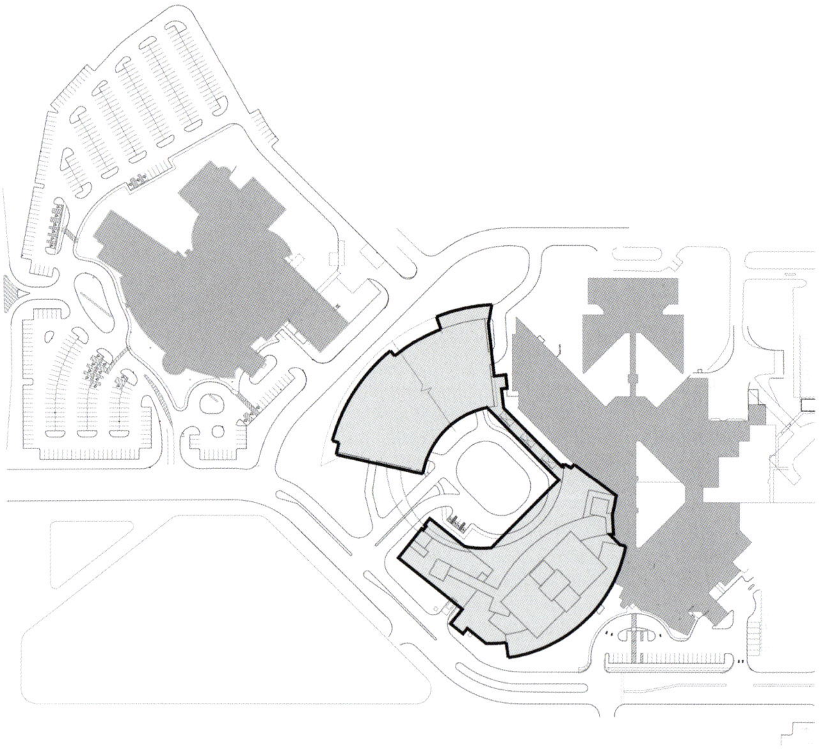

A new six-storey patient tower is the focal point of a recent expansion and renovation of the Community Hospital North in Indianapolis. The project, designed by RTKL, also includes a five-storey medical office building, a six-storey parking garage, and the renovated space in the existing facility. RTKL offered services of architecture, interior architecture and design, medical equipment and planning.

The existing hospital had many entrances, making wayfinding confusing. The designers solved this problem by creating one main entry where all visitors enter a large, two-storey, glass-enclosed gallery that spans the front of the patient tower. The gallery provides easy access to all services and was designed, in addition, to showcase original art by local artists and thereby enhance the hospital's healing environment. Above the gallery, a glass curtain wall admits abundant natural light to patient floors. The airy, art-filled building is designed to reduce the stress, intimidation, and confusion inherent in traditional medical settings.

New and existing areas were equipped with state-of-the-art digital information technology, security and communication systems. RTKL's Healthcare Technologies group provided medical equipment planning for the project.

新建的6层患者楼是印第安纳波利斯北方社区医院翻修工程的焦点。项目还包括一座5层的医疗办公楼，一座6层的停车场和现有设施的修复工程。RTKL事务所提供了建筑、室内设计、医疗设施和规划方面的服务。

医院原来有许多入口，容易造成迷惑。设计师通过在患者楼打造一个大型的双层玻璃门廊解决了这一问题。门廊提供了通往各个服务区的简单通道，同时也展示了当地艺术家的原创作品，提升了医院环境的治愈感。门廊上方的玻璃幕墙保证了病房充足的自然采光。轻盈而充满艺术感的建筑减少了传统医疗设施的压力感、胁迫感和迷惑感。

新建和原有的区域都配备了最新的数码信息技术、安全保障和交流系统。RTLK的医疗技术团队为项目提供了医疗设备规划。

1. Main entrance detail 1.主入口细节
2. Main entrance 2.主入口
3. Roof garden 3.屋顶花园

3

1

1. Main entrance
2. Atrium

1. 主入口
2. 中庭

1. Gallery
2. Support area
3. Neuro/Cardio
4. Admitting
5. Pediatrics
6. Outpatient
7. Endoscopy
8. Emergency Department
9. Imaging
10. Chapel
11. Retail space
12. Offices
13. Physician's office building
14. Existing

1. 画廊
2. 辅助区
3. 神经/心脏科
4. 入住登记
5. 儿科
6. 门诊
7. 内窥镜室
8. 急诊室
9. 成像室
10. 礼拜堂
11. 零售空间
12.办公区
13. 医生办公楼
14. 原有建筑

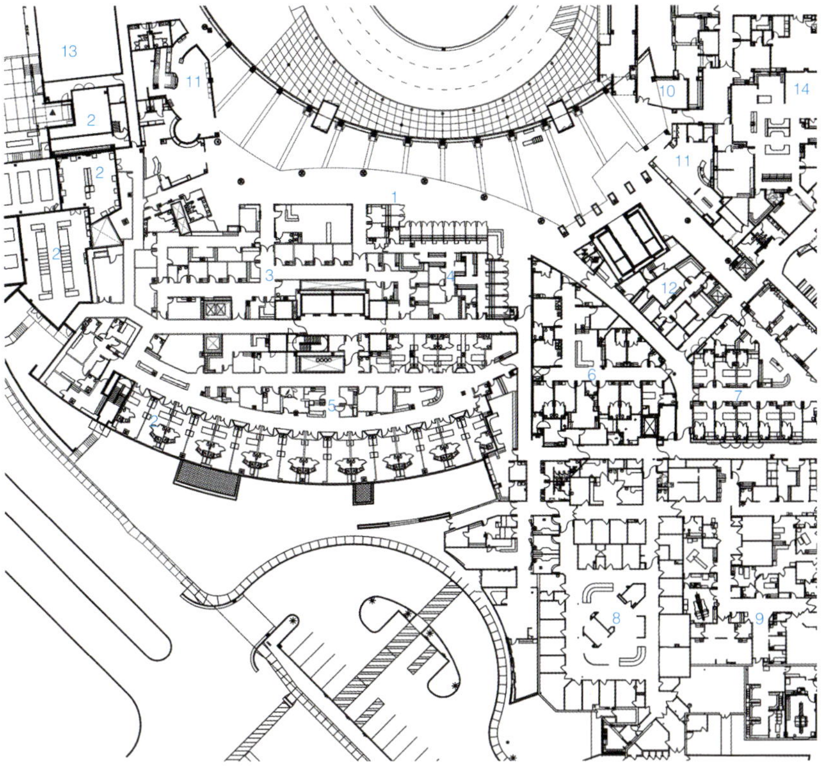

1. A corner inside
2. The gallery showcases original art by local artists
3. Resting area in the central hall
4. The glass structure looks light and modern

1. 室内一角
2. 玻璃回廊展示当地原创的艺术品
3. 中央大厅休息区
4. 玻璃结构轻盈现代

4

1. It is warm and comfortable in the patient room
2. The homelike atmosphere relieve the patients' pressure
3. The office area

1. 病房设计温馨舒适
2. 如家一般的病房减轻了病患的压力
3. 医疗办公区

2

3

道恩医院 Downe Hospital

Location: Downpatrick, UK **Completion date:** 2009 **Designer:** Scott Wilson, in association with Capita **Photographer:** Ory Moore Photography **Area:** 55,000 m^2

项目地点：英国，唐帕特里克 完成时间：2009年 设计师：斯科特・威尔森、卡皮他 摄影师：欧力・莫里摄影公司 面积：55,000平方米

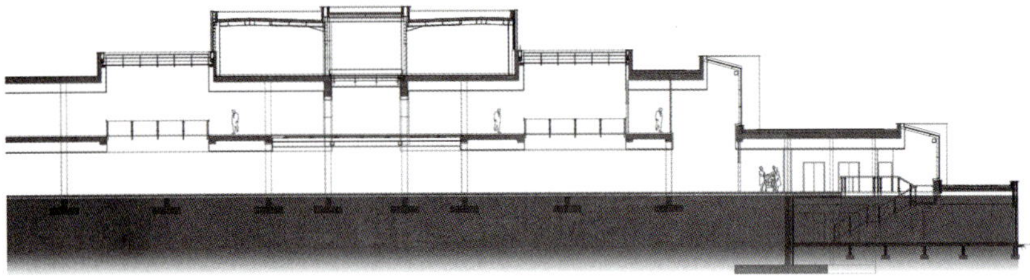

The Downe Hospital site is located within the scenic grounds of the historic Downshire Hospital in the town of Downpatrick, a site which has been associated with providing care to the local community for centuries.

The buildings are designed to a human scale with materials being chosen to compliment the existing hospital and its setting. Whilst providing scenic vistas for all patients, the tired, stepped back form of the buildings blend with the existing formal landscape and provide natural daylight & ventilation throughout.

Scott Wilson established user groups (medical & non-medical) to help in the design of the hospital, including involvement with the local community during the design development to promote ownership of the Hospital, particularly with regard to public spaces. A professional architectural model was created and viewed by the public in the Down area for locals to offer feedback. Several public consultation meetings were held allowing the public to engage first hand with the design team and give feedback on the proposed plans before construction commenced.

Design concepts have been inspired by a community ethos and local art work has been extensively used throughout. Glazed units in the reception include innovative glass art adding vibrancy and assist the level of solar heat in the atrium.

Although the hospital is very contemporary in its design, it is sympathetic to the existing hospital building, integrates well into the scenic landscape and incorporates many historical and local references in the design.

道恩医院位于唐帕特里克风景如画的唐谢尔医院园区之内，近百年来，这里一直为当地社区提供医疗服务。

建筑注重人性尺度，材料的选择上尊重了原有的医院建筑和环境。在为患者提供优美风景的同时，向后错落的建筑与原有的景观环境结合在一起，保证了自然采光和通风。

斯科特・威尔森在医院设计过程中成立了一个使用者小组（兼具医疗与非医疗人员），以体现医院的主人公地位，帮助设计公共空间。专业的建筑模型在道恩地区展示，并接受了本地居民的反馈。在工程开始前，他们召开了几次公共咨询会，让公众与设计团队紧密联系，及时反馈。

医院的设计理念深受社区价值观和当地艺术影响。接待台上极具创意的玻璃艺术为空间增添了活力，也帮助提升了中庭内的热辐射。

尽管医院在设计上十分现代，它仍然能与原有的建筑和谐相处，很好地融入了景观之中，在设计中反映了许多历史和本地特色。

1. The building has an open space in front of it
2. The combination of white and grey merges the building into its surroundings
3. A side view

1. 建筑前方空间开阔
2. 白色与灰色的结合使建筑很好的与周围环境融合
3. 建筑侧面投影

Award:

RIBA Awards (Royal Institute of British Architects), 2010

获奖情况：

2010年英国皇家建筑师协会奖

1

2

3

1. The design emphasises daylight and natural ventilation
2. The orange path is bright and eye-catching
3. The design respects the original architecture style
4. The atrium is surrounded by white architecture

1. 医院设计注重采光和通风
2. 橙色的路面明朗显眼
3. 设计尊重原来的建筑风格
4. 白色建筑环绕中庭景观

4

1

1. The entrance to the hospital
2. A resting area
3. The glass roof provides a good daylight
4. Details of the design

1. 医院入口
2. 休息区
3. 走廊的玻璃顶棚采光良好
4. 设计细节

2

3

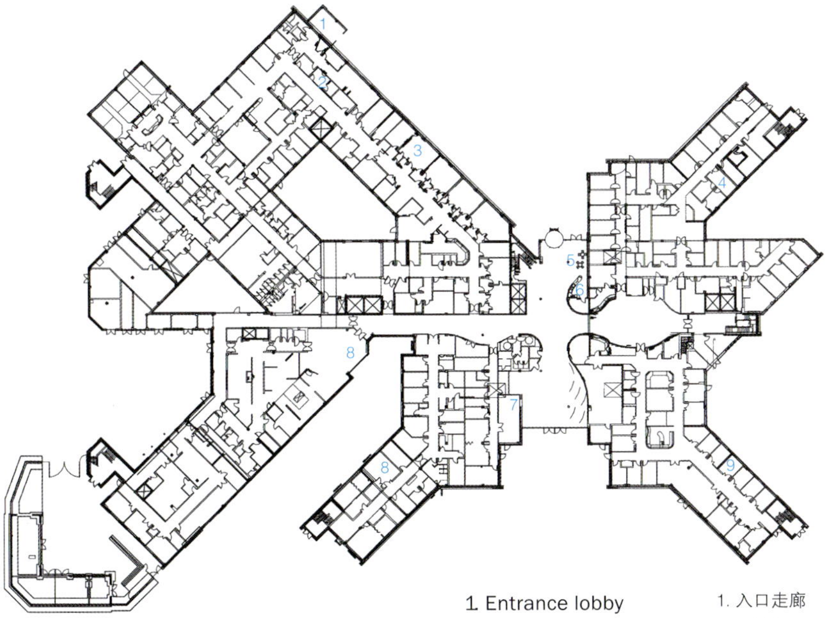

1. Entrance lobby
2. Reception
3. CT room
4. Dental treatment room
5. Waiting area
6. Main reception
7. Library
8. Birthing room
9. Treatment room

1. 入口走廊
2. 接待台
3. 扫描室
4. 牙科治疗室
5. 等候区
6. 总接待处
7. 图书室
8. 分娩室
9. 治疗室

4

泗水尔哈诊所 Erha Clinic Surabaya

Location: Surabaya, Indonesia **Completion date:** 2010 **Designer:** HMP Architects, Heru Mudito Prasetyo, Mirna Listianti **Photographer:** Fernando Gomulya, Sonny Sandjaya **Area:** 3,386 m²

项目地点：印度尼西亚，泗水 完成时间：2010年 设计师：HMP建筑事务所，赫鲁·穆迪托·帕拉斯迪欧、米尔纳·里斯蒂安提 摄影师：费尔南多·古莫尔亚、索尼·桑德加亚 面积：3,386平方米

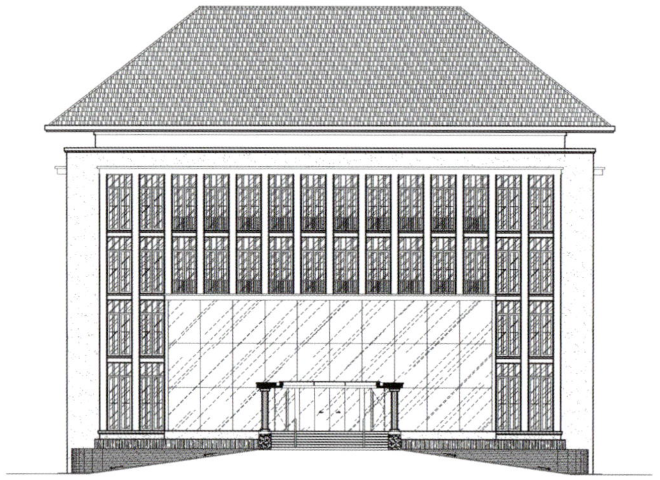

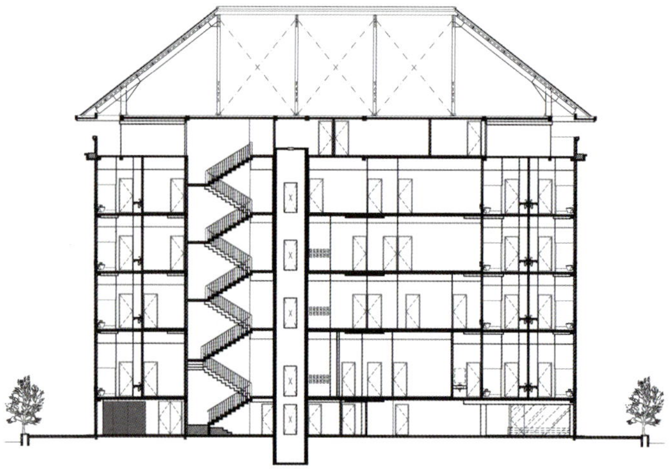

Erha Clinic Surabaya is a skin care clinic, located at one of the big cities in Indonesia – Surabaya. The building is a single mass four-storey building with art deco+modern architecture style. The proportion of the building is divided as building's head, neck and body. Formal and rigid impression of symmetric elements is refined with the presence of the transparent glass box with pattern at the entrance which became one of building point of interests.

The vertical element is very dominant in the building. The concrete thick blade between the windows has aesthetic value to the building and also has a function to reduce solar heat entering the building. The selection of natural stone, concrete, stainless steel, and patterned-glass materials for the exterior was chosen to give the exclusivity and professional impression, considering the building's function as a commercial building which aims to attract visitors. The building mass is placed away from the front site boundary to reduce the bulky impression. The long canopy at the drop-off area which serves as protection from rain and heat also gives the warm and welcoming impression to the visitor of the building. The grass blocks are used in the site, mostly in parking circulation, to allow rain water infiltration into the soil directly.

Natural lighting in the building is achieved through the high windows installed in four sides of the building. The windows provide the natural air ventilation for the building when needed. Two independent circulations are included in the project: the customers' and the doctors' circulation, so that they will not interfere with each other except in the treatment or consultation room.

This clinic completes the variety needs and demands of specific health facilities in Surabaya. Hopefully, the building will add positive values to the area.

尔哈诊所是一家皮肤科诊所，位于印度尼西亚最大的城市之一——泗水。建筑共有4层，采用了装饰艺术与现代相结合的风格。建筑被划分为头、颈、身三部分。透明玻璃盒结构展现了正式和刚性的对称元素，入口处的图案成为了建筑的亮点之一。

建筑注重垂直元素。窗户间的混凝土厚隔断兼具美观和功能性，可以减少楼内的热辐射。自然石材、混凝土、不锈钢和图案玻璃材料的使用让建筑看起来独特而专业，商业化的建筑能吸引更多的人。建筑整体远离场地前端，以免显得笨重。下客区的长棚顶能够遮风避雨，也为来访者营造了一种温暖、欢迎的气氛。停车区的玻璃槽让雨水直接流入地下。

建筑四面高大的窗户保证了室内的自然采光。窗户还为建筑提供了必要的自然通风。项目拥有两个独立的路径：一个为患者，一个为医生准备。在治疗和咨询室他们不会干涉彼此。

诊所满足了特别医疗设施的各种需求，它将会对区域发展起到积极作用。

1. The façade uses glass as main material
2. The entrance
3. The design highlights the vertical elements

1. 建筑外部采用玻璃材料
2. 医院入口
3. 设计重视垂直元素

1

2

KLINIK CABANG UTAMA SURABAYA
erha
klinik spesialis kulit
OUT
3

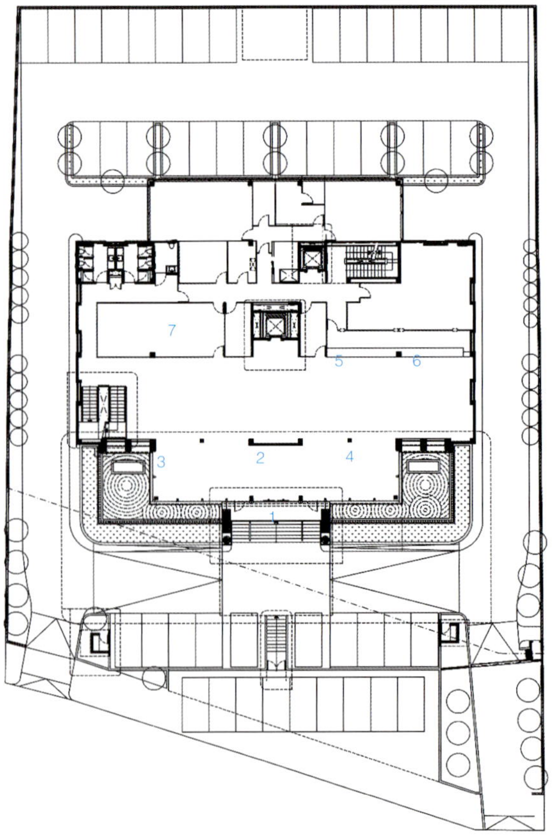

1. The building is quite symmetrical
2. The elegant and luxurious interior design

1. 建筑采取对称风格
2. 室内设计高贵奢华

1. Entrance
2. Information
3. Food corner
4. Waiting area
5. Registration
6. Cashier
7. Treatment

1. 入口
2. 信息中心
3. 食品供应处
4. 等候区
5. 挂号处
6. 收银台
7. 诊疗室

2

1. The interior atrium
2. The unique carved wall
3. White is taken as the main colour in interior design
4. The elegant and comfortable waiting area
5. The reception

1. 室内中庭
2. 雕花墙体独具特色
3. 医院内部采用白色作为主色调
4. 优雅温馨的休息区
5. 接待台

3

4

5

1

2

1. The clean and generous ceiling design
2. The comfortable patient room
3. The treatment area

1. 天花板设计简洁大方
2. 舒适的病房
3. 诊疗室

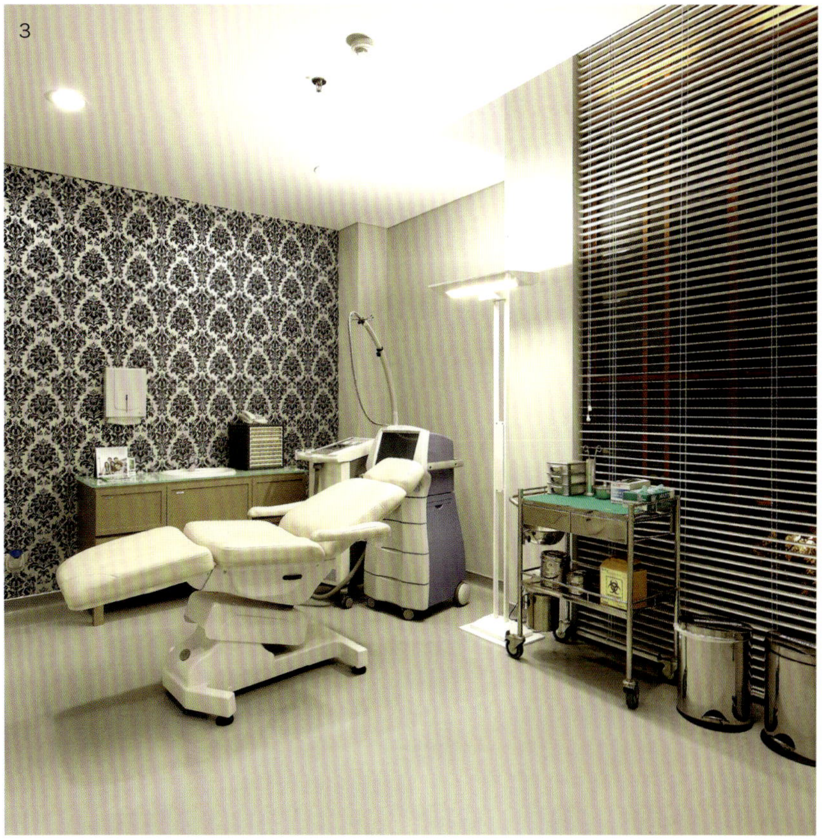
3

金塔夫特医院 Gentofte Hospital

Location: Gentofte, Denmark **Completion date:** 2009 **Designer:** C. F. Møller Architects **Photographer:** Helene Ryttersgaard, Joergen True **Area:** 14,000 m^2

项目地点：丹麦，金塔夫特 完成时间：2009年 设计师：C. F.穆勒建筑事务所 摄影师：海林·雷特斯加尔德、杰尔根·特鲁 面积：14,000平方米

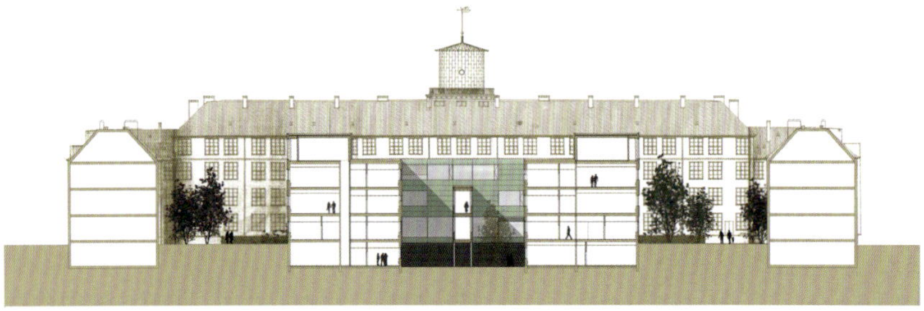

Gentofte Hospital's new facilities are the first phase of C. F. Møller Architects' vision plan for the hospital. The new top-modern treatment wings ensure the future functionality and flexibility of Gentofte's unique historically and urbanistically integrated hospital complex.

The new buildings are a total of 14,000 square metres, and house 24 new high-tech operation theatres, 16 intensive care units, 30 post-operation units, a new main entrance, a new café, several clinical facilities and out-patient departments as well as a new highly automated central sterile supply department. With the inauguration, Gentofte Hospital holds the region's largest operation capacity.

The aim has been to bring the hospital up-to-date with today's technical requirements, and to optimise its logistics. The new layout extends and transforms the urban hospital by inserting a new, modern core in the courtyards of the original figure-eight shaped complex, in the form of two atrium-buildings connected to the original buildings by enclosed walkways. To retain the clear character of the 1920's brick buildings of the complex, the new additions are light and transparent, with satinated glass façades.

The project will also include renovation and modernisation of the existing buildings. All of the hospital's out-patients' departments will be located on the ground floor together with the information and waiting areas, etc., which will improve the overall clarity of the layout and create the best possible patient flow.

On its uppermost storeys, the new area will house the most equipment-heavy rooms, such as the 24 new operating theatres, while on the equivalent storeys, the older buildings will be utilised as in-patient wards. This layout, with the wards located in an eight-shaped pattern around the treatment departments, will minimise the daily transport needs, to the benefit of both patients and staff.

金塔夫特医院的新设施是C. F.穆勒建筑事务所为医院打造的规划的第一阶段。鉴于医院独特的历史价值和区域价值，新的超现代诊疗翼楼将保证未来的功能性和灵活性。

新建部分总面积14,000平方米，拥有24个全新的高科技手术室、16间特护病房、30间术后病房、1个新的主入口、1间新咖啡厅、一些临床设备、门诊部门和1个全新的中央无菌物品供应部。建成后，金塔夫特医院拥有该地区最强的医疗能力。

设计的目的是让医院适应当今的技术要求并且优化它的后勤保障系统。新格局通过在8字形建筑的庭院内引入一个全新的现代核心得以延伸和变形。两座带中庭的建筑通过封闭走道与原建筑连接在一起。为了保证建于1920年的建筑的特点，新建部分轻盈而透明，玻璃表面如缎面般丝滑。

项目还包括对现有建筑的翻新和现代化改造。所有的门诊部门都将设在一楼，与信息和等候区一起。这将提升布局的整体清晰感，并打造最佳的患者流通路径。

顶楼的新区将容纳设备最多的部分，如24间全新的手术区，而旧楼的同等部分将是住院病房。这种格局让病房环绕着8字形的诊疗部门，将最小化日常移动需求，对病患和工作人员都有益。

1. The detail of the building's skin
2. Glass is taken as the main material for the façade
3. The glazed walls ensure the transparency

1. 建筑表皮细节
2. 玻璃作为建筑外墙主要材料
3. 大面积玻璃窗保证了光线的通透性

1

2

3

1. A side view of the building
2. The suspended glass gallery
3. The interior space is open and bright

1. 医院侧面
2. 悬空的玻璃通道
3. 内部空间开阔明亮

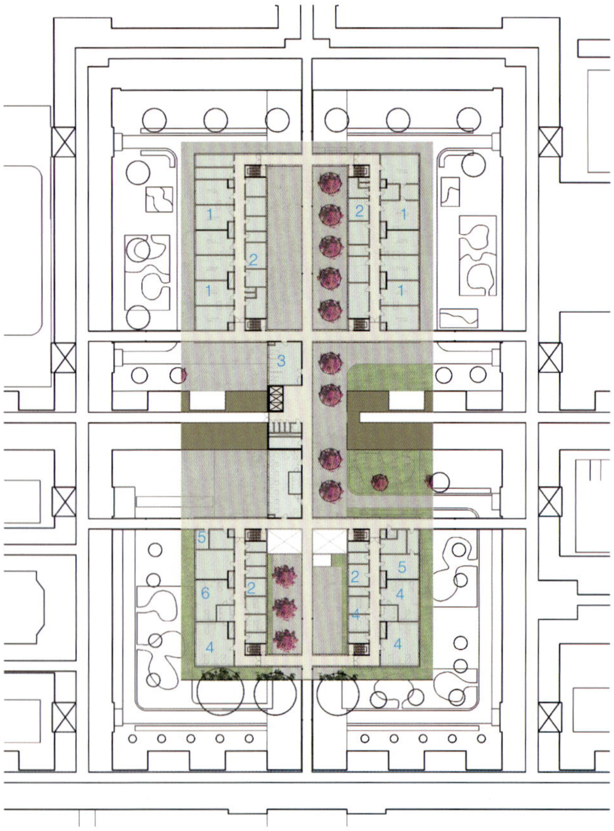

1. Operating room
2. Doctor's office
3. Nurse's office
4. Interior view of intensive care unit
5. Store room
6. Information area

1. 手术室
2. 医生办公室
3. 护士办公室
4. 特护病房内部
5. 储藏室
6. 信息中心

3

Informatio
1

2

Prøvetagning
Klinisk Biokemisk afd.
Café / Kiosk
Diabetesvejleder
Patientvejleder
12 27

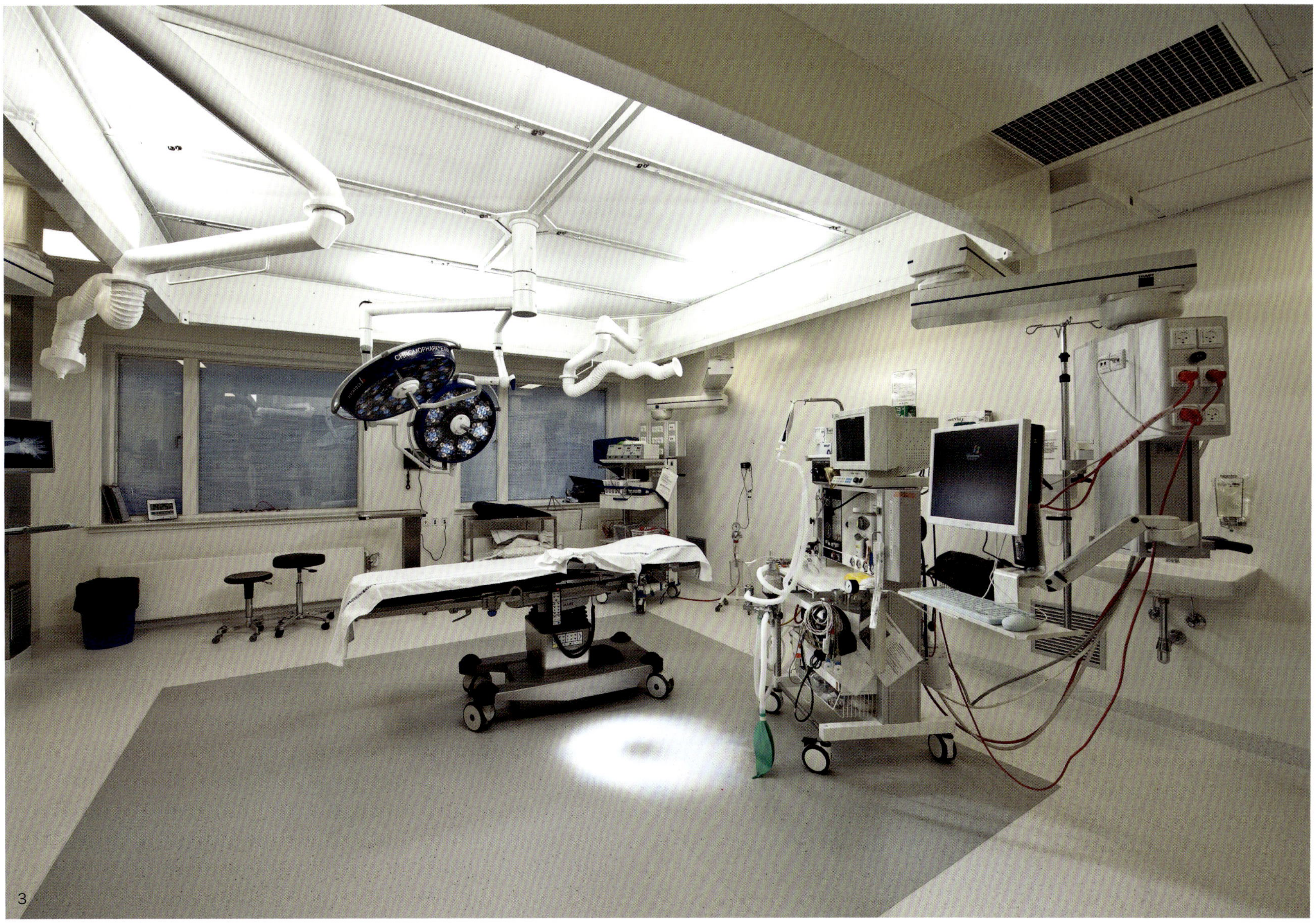
3

1. Waiting area
2. White is the main colour in interior design
3. Treatment room
4. The complicated medical equipment

1. 休息区
2. 白色为室内主要设计色调
3. 医疗室
4. 复杂的医疗设施

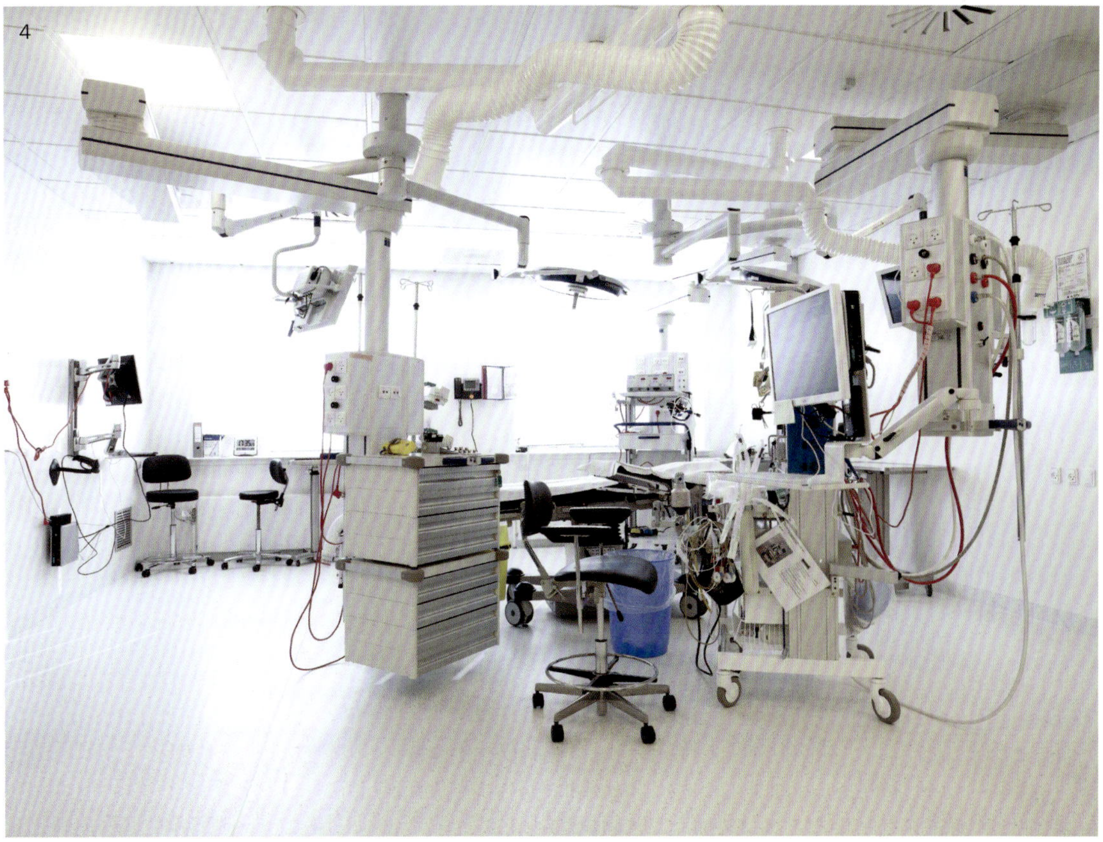
4

吉玛朗伊斯私立医院 Guimarães Private Hospital

Location: Guimarães, Portugal **Completion date:** 2009 **Designer:** Pitágoras Arquitectos **Photographer:** Luis Ferreira Alves Arquivo, Pitágoras Arquitectos **Area:** 14,087 m²

项目地点：葡萄牙，吉玛朗伊斯 完成时间：2009年 设计师：毕达哥拉斯建筑事务所 摄影师：路易・费雷拉・阿尔维斯档案、毕达哥拉斯建筑事务所 面积：14,087平方米

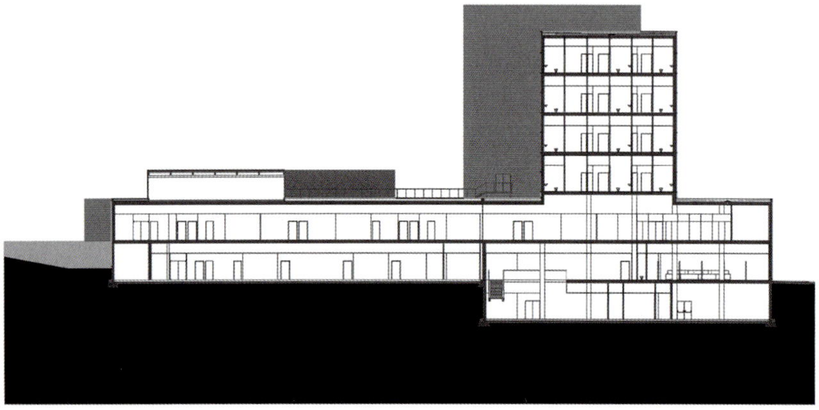

Guimarães Private Hospital is currently located outside of the urban perimeter, near a set of recently developed facilities by the Guimarães City Council – Athletics Track and Swimming Pools Complex – and by individual entities – a new School for primary and secondary education. It is located alongside the chosen Creixomil meadow, made up of land that is part of RAN but in between this land and the sloping land which characterises the Hospital construction area.

The site criteria conditioned the building size and its distribution into two clearly perceptible units, one extensive unit with two floors, developed along the side of the larger land size, and on which overlaps a single and transversal unit which accommodates the internment areas. The aim was for internal and external circulation to be regulated with maximum clarity and that they also correspond to the principle of simplicity which shapes the whole project.

The blocks form a patio and are united by vertical and horizontal circulation components to which public horizontal circulation areas correspond schematically with direct natural lighting, and internal circulation areas for services, or mixed use, mainly artificially lit, which will allow the easy identification of the type of circulation by this simple coding throughout the whole building.

Another priority thought of is that the building should assume a language, scale and character which is eminently public and therefore unique and easily identifiable. This translates into a compact size and clear organisation, with obvious reflections in the covering materials and with a treatment and expression of façades which contributes to that objective – making the Guimarães Private Hospital an individual building, and transforming it into an urban icon.

吉玛朗伊斯私立医院位于城市外围，临近吉玛朗伊斯市议会新建的田径跑道和游泳馆综合建筑以及一所中小学。医院沿着克里克索米尔草场，坐落在一个斜坡之上。

场地限制了建筑的大小，使其被划分成两个简洁明快的部分：一部分是2层楼的巨大模块，另一部分则是与之重叠的细长、横向结构，用作住院区。其目的是将内外部流通路径明确地分开，同时也符合项目简洁的特性。

两个楼体形成了一个天井，并且在垂直和水平方向结合了起来。公共水平循环区域有策略地进行了直接自然采光，内部循环区域则保证了服务、多功能性和人工照明，让人们可以清楚地辨识大楼的各个区域的功能。

设计的另一重点是让建筑拥有自己显著的公共形象、规模和特点，使其容易被辨识。这让建筑尺寸紧凑、组织清晰。建筑的外墙让吉玛朗伊斯私立医院看起来独立的同时也是一个城市新地标。

1. The hospital is surrounded by trees
2. Side elevation of the building
3. The building is divided into two units, vertically and horizontally

1. 树丛中掩映的医院
2. 建筑侧立面
3. 医院主体由纵横的两大模块组成

1

2

3

1

2

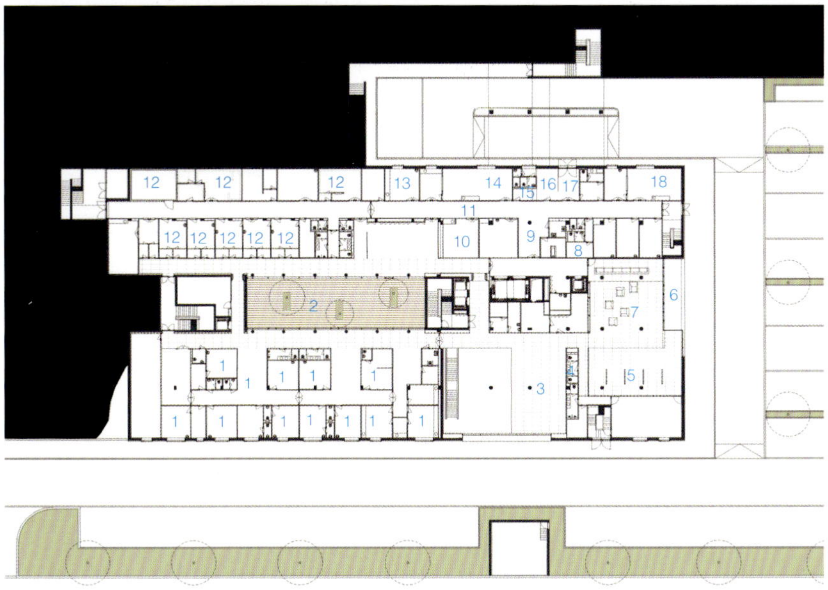

1. Medical office
2. Terrace
3. Sitting room
4. Toilet
5. Attendance
6. Entrance
7. Lobby
8. Toilet
9. Emergency waiting room of patience
10. Treatment room
11. Hall of internal emergency service
12. Exam room
13. Plaster room
14. Nursing room
15. Toilet
16. Waiting room
17. Entry of emergency service
18. Pediatrics room

1. 医疗办公室
2. 露台
3. 休息室
4. 洗手间
5. 考勤室
6. 入口
7. 大堂
8. 洗手间
9. 急诊候诊室
10. 治疗室
11. 内部急诊服务厅
12. 检查室
13. 石膏室
14. 护理室
15. 洗手间
16. 等候室
17. 急诊服务入口
18. 儿科

1. The windows of different sizes add lively atmosphere to the architecture
2. The architecture has a clean and simple style

1. 形状不一的窗子给建筑增添了活跃的气氛
2. 建筑风格简洁干练

1

2

1. The interior natural landscape
2. The pathway under the building's volume
3. The French windows supply efficient daylight
4. The clean and simple corridor design
5. The entrance to the medical area

1. 建筑内部的自然景观
2. 医院主体下的通道
3. 大落地窗使室内采光良好
4. 走廊设计简洁
5. 医疗区入口

3

4

5

1

2

1. Green sofas add some peaceful atmosphere
2. Inside the patient room
3. The central hall is quite spacious

1. 绿色沙发增添了平静祥和的氛围
2. 病房内部
3. 中央大厅十分开阔

3

海德威格舒和医院翻修和扩建工程

Remodelation and Extension of Hedwigshöhe Hospital

Location: Berlin, Germany **Completion date:** 2008 **Designer:** Huber Staudt Architekten BDA, Manuel Brullet and Albert de Pineda **Photographer:** Werner Huthmacher, Jordi Bernadó, Huber Staudt Architekten BDA

项目地点：德国，柏林 完成时间：2008年 设计师：胡贝尔·施托德建筑事务所、曼纽尔·布鲁利特、艾伯特·德·皮纳达 摄影师：维尔纳·休斯曼彻尔、乔迪·博尔纳多、胡贝尔·施托德建筑事务所

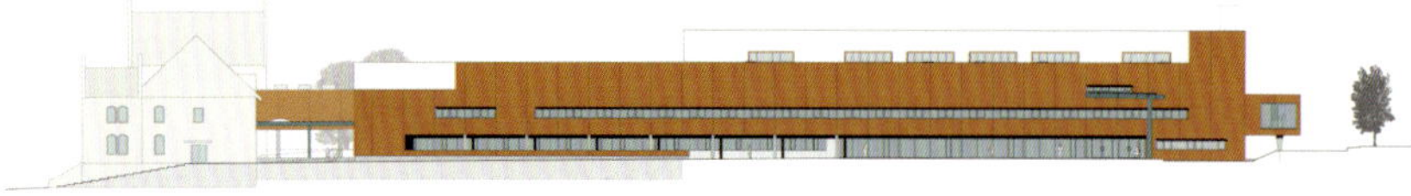

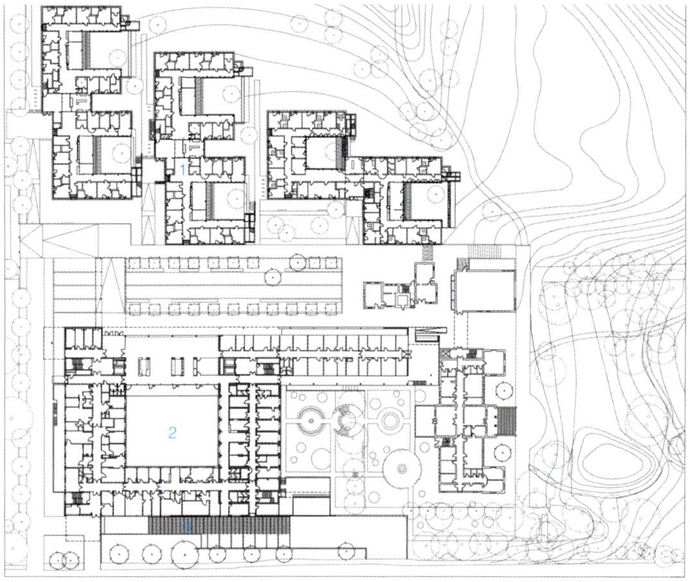

1. Inpatient
2. Courtyard
3. Corridor

1. 住院部
2. 中庭
3. 走廊

The intervention is based on the remodelling and extension of the existing hospital and on the design of the psychiatric units. In the open dialogue between the dense urban pattern and the spacious landscape, the new Hedwigshöhe Hospital is the turning point for Falkenberg, a small garden city in Berlin's southeast suburbs. To contrast the existing structure, the two-storey pavilions of the psychiatric ward are playfully juxtaposed; the result is a small hospital city where the parataxis between old and new is evident.

The new psychiatric department is located north, on the slope of Flakenberg hill. Patients are welcomed in paired pavilions, each set as to create a slight crenelation towards the surrounding open landscape. The two levels follow the scale of close-by homes and their patios reach for a view over the landscape; the façades of the courts are wood and glass and wooden battens characterise the hospital that is defined by the contrast between white plaster and horizontal wooden fit-ups. Along the external perimeter, the square windows are divided in three parts and slightly project from the wall.

The hospital is on five levels. On basement level are therapy units, kitchen, cafeteria, ancillary services, and staff locker-rooms. On ground floor are: entrance hall, cafeteria, emergency units, laboratories, radiology, MRT, CT, functional diagnosis, endoscopy, medical services, psychiatry and three 24-bed psychiatric wards. In the villa is the archive, day-hospital and psychiatric visiting rooms, in the tower is the obituary and the chapel. On hospital's first floor are two 32-bed interdisciplinary wards, 8-bed multidisciplinary intensive care units, two 24-bed psychiatric wards, areas for occupational therapy, staff changing rooms. In the villa are meds offices while administration is in the tower. On hospital's second and third floor are other two interdisciplinary wards and on the last floor of the tower other meds offices. The particular feature of this hospital is in the floor plan, that is so discordant from the classical typological configurations. The schemes seems to support a positive contrast between the orderly layout of the existing structures and the vivacious layout of the psychiatric courts; the new sanitary structure establishes a conscious relationship with the landscape, the city and the final user.

工程以翻修和扩建原有医院并设计精神科病房为基础。在密集的城市网格和宽阔的景观之间的对话里，新海德威格舒和医院是法尔肯堡市——一个柏林东南郊区的小型花园城市的转折点。为了与原有结构进行对比，两层楼的精神科病房与其巧妙地并列在一起，二者新旧的对比十分明显。

新的精神科位于北侧法尔肯堡山的山坡上。成对的凉亭欢迎着病患，在开阔的景观中形成了开口。两层的建筑与附近的主宅相似，它们的天井与景观相连接。建筑的外墙由玻璃和木板组成，木板与白色石膏形成了香茗的对比。正方形窗户被分为3部分，略微从墙面上凸出。

医院共分5层。地下室是治疗室、厨房、餐厅、附属设施和员工更衣室。一楼是入口大厅、餐厅、急救室、实验室、放射线科、CT室、功能诊断室、内窥镜检查室、医疗服务、精神科和3间24个床位的精神病房。别墅里是档案室、日间医院和精神病会客室。塔楼里是讣告室和礼拜堂。医院的二楼是2间32个床位的跨学科病房、8床位的跨学科特护病房、2间24个床位的精神病房、职业心理疗法区和员工更衣室。别墅里是医疗数据办公室，塔楼里是行政办公室。医院的三楼和四楼是另两个跨学科病房。塔楼的顶层是一些其他的医疗数据办公室。医院设计的特别之处在于平面规划，与典型的配置大不相同。设计方案与有序的原有结构形成了积极的对比，显得更加活泼。新的医疗设施与景观、城市和最终使用者建立了鲜明的关系。

1. The architecture mixed well with the landscape
2. The main volume of the building is white
3. The façade is made of wood structure

1. 建筑与自然景观完美融合
2. 白色的建筑主体
3. 木制结构构成了建筑外墙

Award:

Honourable Mention BDA-Prize 2009 (Federation of German Architects)

获奖情况：

2009德国建筑师联盟荣誉奖

1

2

3

1. Wood makes a great contrast with white plaster
2. Wood pathway
3. The corridor is red and white
4. Waiting area

1. 木板材料与白色石膏形成鲜明对比
2. 木制通道
3. 红白相间的走廊
4. 休息区

CUF医院 Hospital CUF

Location: Porto, Portugal **Completion date:** 2009 **Designer:** Manuel Ventura & Associados Arquitectos **Photographer:** António Teixeira **Area:** 7,500 m^2

项目地点：葡萄牙，波尔图 完成时间：2009年 设计师：曼纽尔·文图拉建筑事务所 摄影师：安东尼奥·泰克瑟拉 面积：7,500平方米

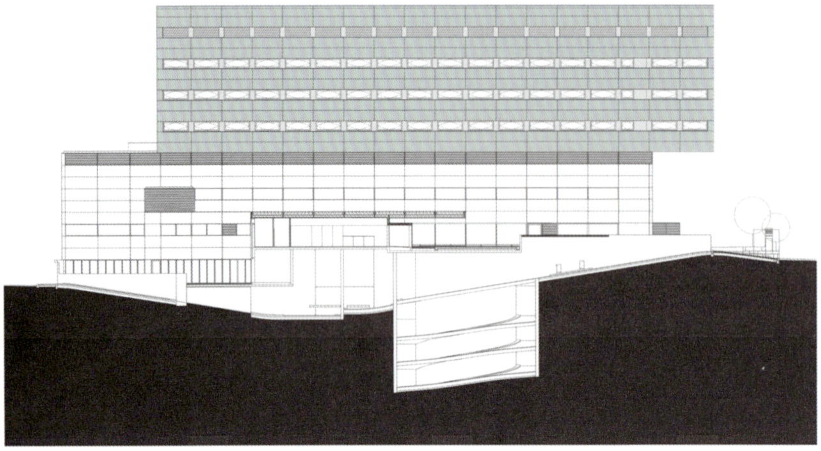

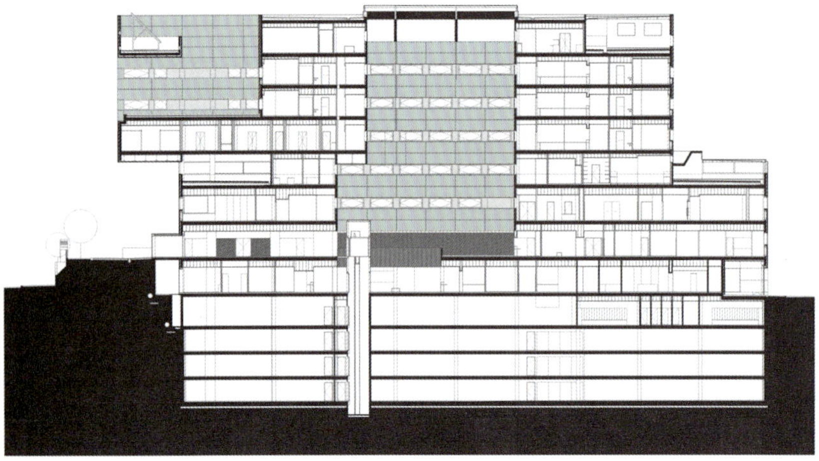

The building is very compact and consists of two overlapping volumetric bodies. The upper volume is coated with glass and steel and is disjointed over a swing of eight metres on the more opaque body, made of concrete, which forms the base. Thereby it protects the entrance and moving away from the subsequent area.

The coating materials of the upper body provide brightness effects and hues of light that enrich their reading and emphasise the volumetric differentiation inherent to the design of the building. The upper volume mainly houses the parts of the programme for use of the general public, namely the internment, while the more technical areas are located at the base, including "triage" assessment and medical and surgical operations.

The foyer is a constructive emptiness that lets in natural light for a far-reaching part of the programme located inwards in the more remote areas from the façades. It is a space that allows a global view of the building, as an internal square.

The structure of the circulations is done through columns of vertical access with stairs and lifts located in the four corners of the large foyer. Staff and goods entrance are through the back and public access is near the entrance.

The waiting areas of outpatients appointments are located in the foyer base, in the floor level -1, which is accessed from the reception on escalators or lifts. The atmosphere created with textures and warm colours materials contrast with the steel and glass coating of the foyer elevations in the upper floors.

The hospital also has a car park with sufficient capacity to support all the flux generated by the hospital traffic.

建筑结构十分紧凑，由两个错落的结构叠加而成。上方的结构是玻璃和钢板外墙，下方的底座结构相对不透明，由混凝土建成。这样一来，下层结构便保护了入口，并且与后方区域相隔较远，上层结构的表皮材料提供了亮度效果和轻快的色调，增强了自身的辨识度，也强调了建筑设计的空间差异。上层结构主要包含公共功能，也就是收容设施；而更具技术性的区域则设在底层，包括伤患分类评估、内科和手术室。

门厅结构空旷，让自然光可以到达大楼的深处、远离外墙的区域。从这里可以纵观全楼，它相当于一个内部广场。

路径结构通过门厅四角垂直通道的楼梯和电梯组成。员工和物品入口设在后部，而公共入口则设在前门。

门诊等候区设在门厅下方的地下一层，可以从接待处的自动扶梯和电梯进入。材料的纹理和温暖的色彩与上层结构的钢铁玻璃外墙形成了鲜明的对比。

医院还拥有一个足够容纳来往于医院的车辆的停车场。

1. The hospital is surrounded by trees
2. The building is divided into two volumes
3. The façade of the hospital

1. 树木掩映中的医院
2. 建筑主体由两大模块构成
3. 医院外立面

Award:
World Architecture Festival, 2010, Shortlist

获奖情况：
2010年世界建筑节入围奖

1

2

3

1

2

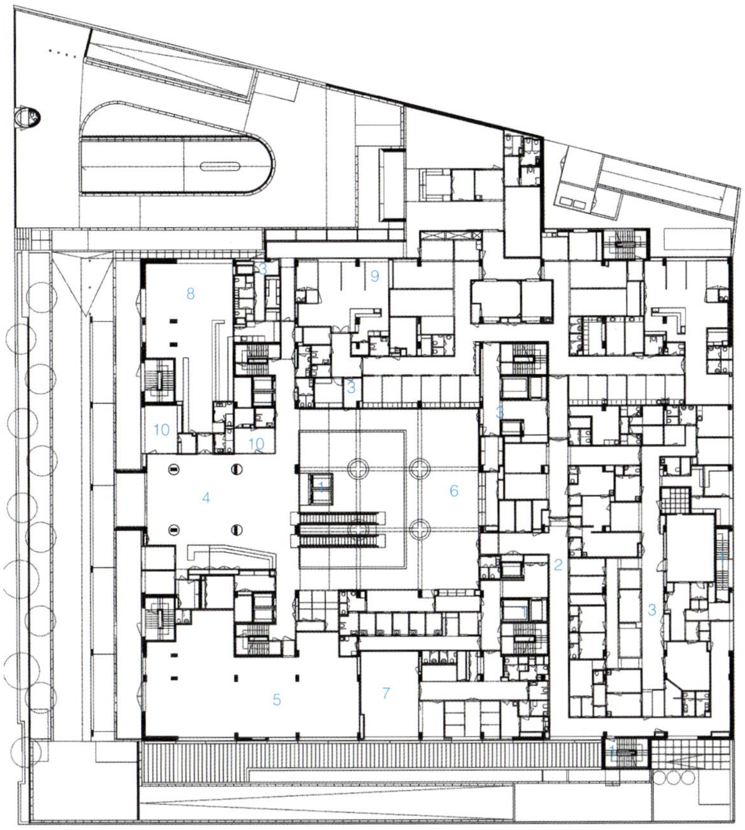

1. The interior space is open and full of daylight
2. The design has a futuristic sense

1. 建筑内部空间开阔，采光良好
2. 室内设计具有未来感

1. Staircase and elevator
2. Public corridor
3. Employees corridor
4. Reception
5. Empty area
6. Diagnostic imaging
7. Laboratory medicine
8. Coffee shop
9. Emergencies and urgent care unit
10. Store

1. 楼梯和电梯
2. 公共走廊
3. 员工走廊
4. 接待处
5. 空置区域
6. 诊断影像室
7. 医学实验室
8. 咖啡厅
9. 急诊室与紧急监护病房
10. 商店

1

2

3

4

1. The atrium is surrounded by layers of floors
2. A glance of the corridor
3. Glass and steel are the main structural materials
4. The white interior space
5. The clean pathway

1. 层层楼梯环绕的中央大厅
2. 走廊一瞥
3. 玻璃和钢板为主要构造材料
4. 纯白色调的室内空间
5. 整洁的通道

5

纳瓦拉医院 Hospital Navarra

Location: Navarra, Spain **Completion date:** 2008 **Designer:** Enrique Rodríguez Antón **Photographer:** Alfonso Percaz **Area:** 7,500 m^2

项目地点：西班牙，纳瓦拉 完成时间：2008年 设计师：恩里克・罗德里格斯・安东 摄影师：阿方索・珀尔凯兹 面积：7,500平方米

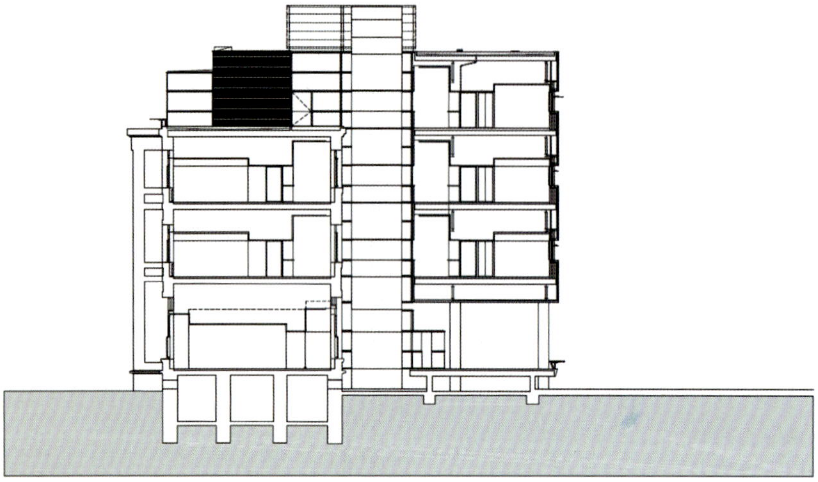

The Pavilion D of the Hospital Navarra, is one of the four major central pavilions, now interconnected with the central building Radiology interposed between the B and D and pharmacy.

The new programme should be implemented in a resounding, by alleviating surface, to use the new function properly. Thus, the strategy does not only require prosthetic regain the old functionality, but the new required, for which inevitably requires dressing pathway as drastic solution.

The strategy seeks prosthetic disguise dressing was not under any circumstances the result of the intervention, nor conceal the new use, let alone mimic the body it serves: it manifests its mission prosthetic explaining their new role. The architects need to solve the following issues: surface extension, provision of a strategy that seeks a high degree of flexibility, reorganisation flow horizontal-vertical movement, New connections, and Re-structuring of existing connections.

The new structure is a parallel to the existing flag, leaving a courtyard between the original and the extension flag. The new pill is a parallelepiped built on three levels, leaving the ground floor (creating in it a covered porch for ambulance access), whose mission is homogeneous container-room areas. The orientation of the rooms is opposed to the original flag and creates a glazed interior corridor that contains the main circulation. Only the area for Coronary Care Unit, to require a larger bay, represents an expansion by suturing the two volumes covering the first floor patio.

纳瓦拉医院D楼是四个中心楼之一，与主楼放射科相连。主楼则环绕在B楼、D楼和药剂楼之间。

新项目的实施十分彻底，要求建筑外观缓和、让人放松，新功能得到适当的安置。因此，项目设计不仅要修复原有的功能，还要满足项目的新需求，对通道进行彻底的装饰。

设计力求让外观设计在不影响建筑的前提下展现新功能，通过新外观反应建筑的新角色。建筑师需要解决以下问题：表面扩展、打造一个高灵活度的方案、重新组织水平和垂直路径、建立新的连接点和重塑原来的连接点。

新结构平行于原来的外墙，在二者之间形成了一个庭院。新外壳是一个平行六面体，覆盖了上面的三层，一楼被留了下来（外壳形成了门廊可以方便救护车进出）。一楼是均匀的集装箱式空间。房间的朝向与原来相反，所形成了玻璃走廊是主要流通路径。由于冠心病监护病房区需要更大的空间，二楼天井的两个空间被结合在一起，达成了它的扩建。

1. The façade of the hospital
2. The design emphasises vertical elements
3. The new volume is a parallelepiped

1. 医院外立面
2. 建筑设计重视垂直元素
3. 新墙体是一个平行六面体

1

2

3

1

1. A side view of the building
2. The rooftop structure

1. 建筑侧面投影
2. 医院顶部构造

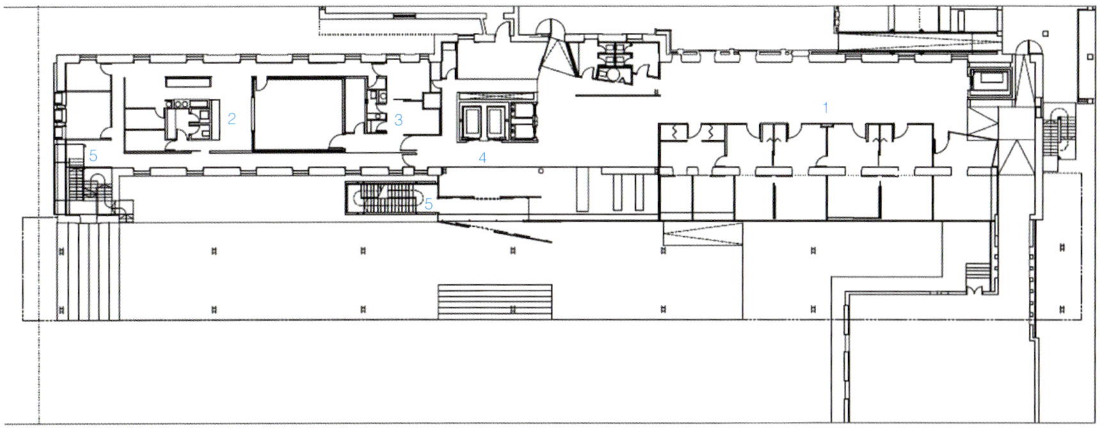

1. Surgical department
2. Office
3. Lavatory
4. Lifts
5. Stairs

1. 手术室
2. 办公室
3. 洗手间
4. 电梯
5. 楼梯

1

2

3

4

5

1. The new construction is light and flexible
2. The new and old walls are parallel
3. The ramp
4. Waiting area
5. The clean corridor

1. 新结构轻盈灵活
2. 新旧墙体之间互相平行
3. 斜坡走道
4. 休息区
5. 整洁的走廊

克拉根福医院 Hospital Klagenfurt am Wörthersee

Location: Klagenfurt, Austria **Completion date:** 2010 **Designer:** DFA Dietmar Feichtinger Architects, "P" ZT GmbH Architekten + Ingenieure, Müller & Klinger/Architects Collective AC ZT-GmbH **Photographer:** Hertha Hurnaus, Wolfgang Thaler, Gisela Erlacher **Area:** 120,000 m^2

项目地点：奥地利，克拉根福 完成时间：2010年 设计师：DFA建筑事务所、"P" ZT建筑工程事务所、穆勒&克林格尔设计公司 摄影师：赫沙・休毛斯、沃尔夫冈・泰勒、吉塞拉・厄尔拉切 面积：120,000平方米

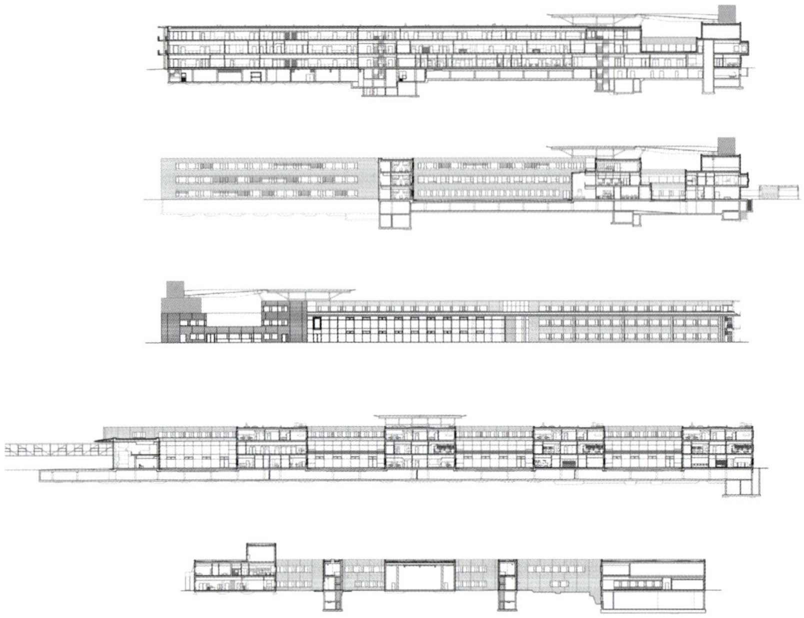

The flat, low-lying structure merges easily with the low and heterogeneous urban surroundings of Klagenfurt. At the north side of the complex lays the floodplain of the river Glan. This scenic green space provides a filter between the urban space of the city and the hospital grounds. The horizontal division of the façade and the tiered structure of the form combine to further integrate the complex into the landscape. The comb-like organisation of the buildings provides a strong visual relationship to the surrounding green areas for both the accommodation and medical wards. Thus the hospital rooms, access halls and waiting areas all become sun-drenched and comfortable health-stimulating light therapy rooms. This tangible connection to the natural surrounds is an integral part of the new Hospital Klagenfurt.

The direct link between the patients and staff to the lush riverside and planted courtyards is an integral part of the architectural concept. Each garden courtyard between the different programmes of the hospital is differentiated by the specific design and colour of its diverse vegetation. This means that these green spaces can also be used as references for orientation within the complex. The ground cover and vegetation is given ample soil and space to grow and mature in the coming years as the landscape becomes increasingly more natural.

The primary circulation of the building is from the north-east corner to the entrance hall in the centre of the hospital grounds. The main entrances are all easily reached by bike or public transit from the city centre. Western access to the Surgical-Medical Centre is also connected to the parking garage and the open visitor's car park.

To take advantage of natural lighting, the programmed spaces are east-west oriented. The specialty medical and outpatient departments are organised along the central patient corridor which forms the structural and functional spine of the facility. The wards and medical units are accessed via the two main east-west corridors and through the entrance hall. North-south corridors run along the courtyards and connect the wards with the central clinic building to the south.

Each ward is equipped with a compact unit of two nursing stations which are centrally located with easy access to the supply/disposal system. All hospital rooms and lounges are oriented towards the courtyards to take advantage of natural light. The intensive care wards are located in the immediate vicinity of the operating area located on the first floor. These wards are connected to the operating rooms by internal corridors at the nursing stations.

扁平的建筑结构简单地融入了克拉根福低矮多变的城市建筑群。医院北侧是格兰河的泄洪平原，这片景观绿地在城市空间和医院之间起到了过渡作用。建筑外墙的水平划分和层叠结构进一步使其与景观融为一体。住宿和内科病房的梳形结构与周边的绿地形成了强烈的视觉联系。医院病房、大厅和等候区都沐浴在阳光里，可以提供舒适健康的光照疗法。与自然景观如此紧密的联系是克拉根福医院最主要的特征之一。

患者和工作人员与青葱的河畔和庭院的直接联系是医院设计的重要环节。不同区域之间的花园庭院拥有不同的设计，并且通过不同种类的植物形成了不同的色彩。这意味着这些花园也可以作为医院的导向标志。地被植物和草木被给予充足的土壤和空间来生长，以后这些景观将会变得更加自然。

建筑内部的主要通道从东北角延伸到中央大厅。从市中心可以方便地骑车或乘坐公交车到达医院的各个主入口。外科医疗中心的西入口与车库和开放式外来停车场相连。

为了保证自然采光，区域空间全部采用东西朝向。特色医疗和门诊部门沿着项目结构和功能的中心走廊设置。病房和医疗室可以通过东西向的走廊和入口大厅进入。南北向的走廊沿着庭院而设，将病房与南侧的中心大楼连接起来。

每间病房的中心都配备了两个护理台，便于患者接近供给系统。所有的房间和休息室都朝向庭院，以获取最佳的自然采光。特护病房紧邻二楼的手术区，通过护理台的内部走廊与手术室相连。

1. The location has a convenient transportation
2. The high winding corridor is in an imposing manner
3. A night view of the building

1. 建筑周围交通便利
2. 高大的回廊气势恢宏
3. 夜景中的建筑

1

2

3

1

2

3

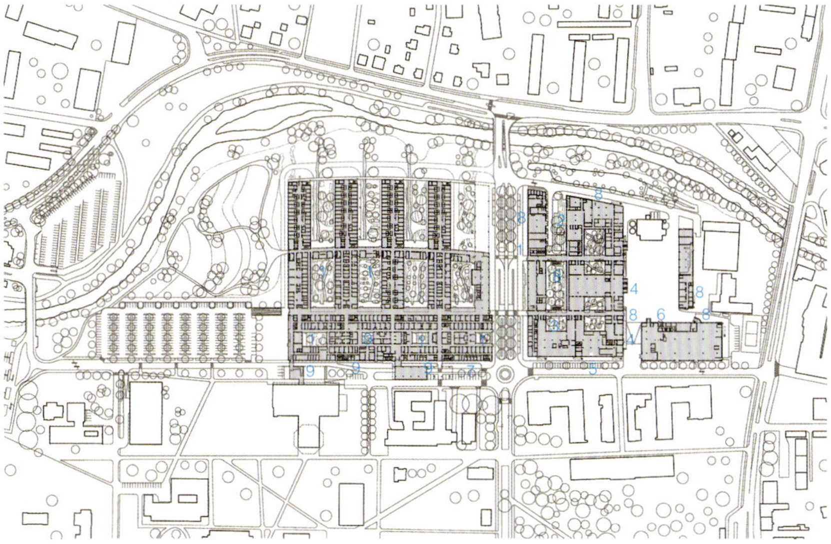

1. Office
2. Cloakroom-staff
3. Laboratories
4. Logistics
5. Pharmacy
6. Laundry
7. Diagnostics
8. Technology
9. Surgery

1. 办公室
2. 员工更衣室
3. 实验室
4. 后勤部
5. 药房
6. 洗衣房
7. 诊断室
8. 机械室
9. 手术室

1. The architecture applies stack structure
2. The glazed volume is clear and bright
3. The expanded mid-sky corridor

1. 建筑采用层叠结构
2. 玻璃主体通透明亮
3. 延伸的空中通道

1

2

3

1. The use of warm colours makes the patients feel intimate
2. Waiting area
3. The open corridor

1. 暖色调的使用使患者产生亲切感
2. 休息区
3. 开阔的走廊

1

1. The clean pathway
2. The interior design is clean and generous
3. The design is quite artistic

1. 整洁的通道
2. 简洁大方的室内布置
3. 设计充满艺术感

鲁兹医院 Luz Hospital

Location: Lisbon, Portugal **Completion date:** 2007 **Designer:** RISCO Architects **Photographer:** FG + SG - Fotograa de Arquitectura **Area:** 110,000 m^2

项目地点：葡萄牙，里斯本 完成时间：2007年 设计师：RISCO建筑事务所 摄影师：FG + SG建筑摄影公司 面积：110,000平方米

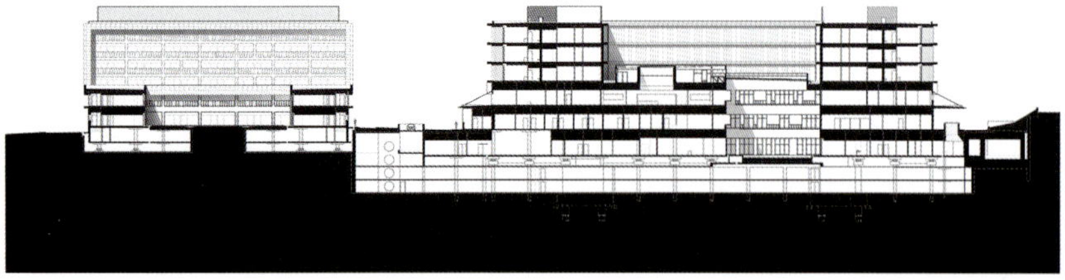

The Luz Hospital is a relatively new and very modern medical facility that is located in Lisbon, Portugal. Those who are planning to travel to Lisbon for pleasure would be prudent to take note of the location of this hospital. No one ever knows when good medical care will be necessary, and having to seek it out after the fact is very stressful. The Luz Hospital provides both Outpatient and Inpatient care and treatment for people of all ages.

The Luz Hospital is one of the largest hospital complexes in Lisbon. Located near the Luz Stadium and the Colombo Shopping Centre, it is bordered by a number of major road arteries leading north from the city.

The small size of the site meant that the designers had to opt for a compact solution with a single structure, which is crossed by an internal road with links to the Avenida Lusíada and the Pontinha trunk road. The internal road separates the two main units within the complex – the Hospital and the Seniors Residence. The Hospital has 193 rooms, while the Seniors Residence contains 118 apartments.

The H-shape layout of the Hospital and the rectangular floor plan of the Residence allowed the designers to include gardens within the site. The need for natural light was thus a major factor in the design.

The health complex is clearly laid out in terms of what goes where. The internal traffic network orders the different functional areas of the complex and relates them to one another, thereby helping to make it easy for people to find their way around. The various internal routes clearly separate in-patients from the general public, and foster the privacy, comfort and confidentiality of both in and outpatients.

The fact that the site is located over two metro tunnels meant that the complex's foundations include anti-vibration supports.

鲁兹医院是一座位于葡萄牙里斯本相对新而现代的医疗设施。来到里斯本的游客最好记住这所医院，以备不时之需。鲁兹医院为各个年龄段的患者提供门诊和住院服务。

作为里斯本最大的医院之一，鲁兹医院临近鲁兹体育场和科伦坡商业中心，周边由许多通往城市北部的交通干道。

场地限制让设计师必须选择一个紧凑的解决方案来打造单体建筑，园区内道路连接着阿维尼达・露丝艾达和普丁哈公路，也分开了医院和老年公寓。医院共有193间房，而老年公寓有118套公寓。

医院的H形布局和老年公寓的矩形平面规划让设计师可以在场地中设置花园。自然采光也是设计的重要因素之一。

医院的布局十分清晰。室内交通网络调整了不同的功能区，并将它们联系起来，从而帮助了人们找到自己的目的地。各种内部路径将住院部门和公共区域分隔开来，为住院和门诊病人打造了私密、舒适而秘密的环境。

由于场地位于两个地铁站之间，医院的地基进行了减震支持。

1. The architecture has a simple and clean style
2. A side view of the hospital
3. Details of the architecture

1. 建筑风格简洁
2. 医院侧面
3. 建筑细节

1

2

3

1. The design is compact and generous
2. There is a landscape garden in the centre

1. 设计紧凑大方
2. 场地中央设有景观花园

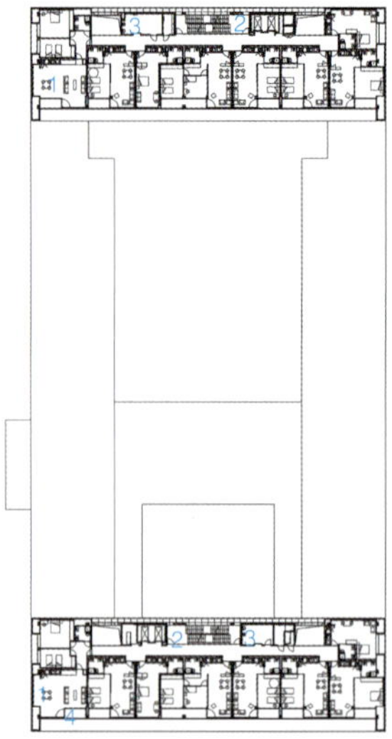

1. Apartments
2. Vertical access
3. Support
4. Balcony/terrace

1. 公寓
2. 垂直入口
3. 辅助设施
4. 阳台/露台

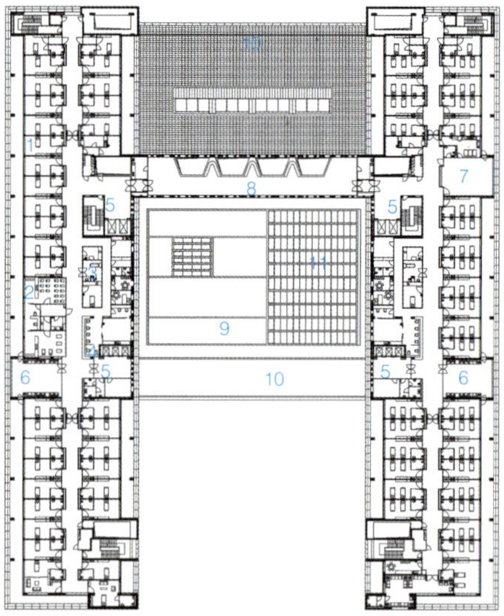

1. Hospital room (with bath room)/suite
2. Nursery
3. Hospital support (treatment room, helped bath, deposit material)
4. Reception
5. Vertical Acess/hall
6. Waiting room/living-room
7. Dining-room
8. Circulation/living
9. Technical area
10. Terrace
11. Skylight (interior courtyard)

1. 病房（带浴室）/套房
2. 护理室
3. 医院辅助设施（治疗室；辅助洗浴；寄存材料）
4. 接待处
5. 垂直入口/大厅
6. 等候室/起居室
7. 餐厅
8. 通道/起居空间
9. 技术区
10. 露台
11. 天窗（室内庭院）

2

1

2

1. The interior design is simple and open
2. The building has a clear interior circulation layout
3. The corridor in white has abundant daylight

1. 建筑内部简洁开阔
2. 室内交通网布局清晰
3. 纯白色调的走廊采光良好

3

新维多利亚医院 New Victoria Hospital

Location: Glasgow, UK **Completion date:** 2009 **Designer:** HLM Architects **Photographer:** John Cooper **Area:** 30,860 m^2

项目地点：英国，格拉斯哥 完成时间：2009年 设计师：HLM建筑事务所 摄影师：约翰·库伯 面积：30,860平方米

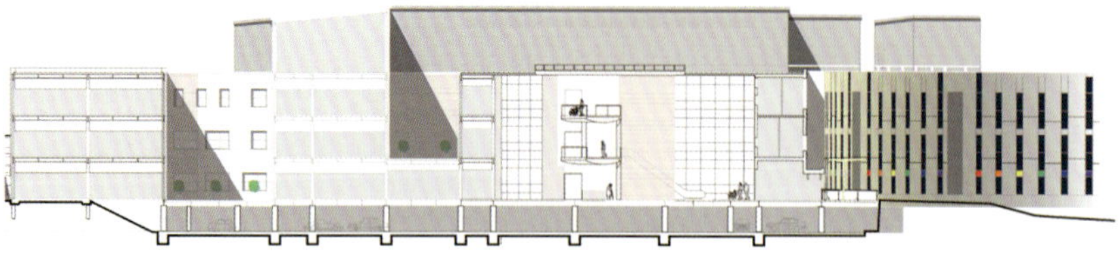

The new Victoria Hospital treats around 400,000 patients annually by offering integrated diagnostic and treatment services including outpatient clinics, day surgery, rehabilitation, and specialised emergency services, thereby introducing an ambulatory model of one-stop care in South Glasgow based on the following key principles: increased accessibility to ambulatory care for the local community, prevention of emergency work adversely affecting the efficient delivery of elective care, improved quality of patient focused care, streamlining of the patient journey to provide a one-stop service, achieving compatibility between healthcare delivery and technology in a fit for purpose building.

Within the context of this clinical brief, the new building has been designed to respond positively to its unique urban setting. To reduce the impact of surrounding traffic noise and ensure the comfort of patients and staff, the hospital has been conceived as a highly insulated “sealed box” incorporating full environmental and comfort cooling.

The building has been configured to allow maximum retention of existing mature trees on the site. The elevations respond to the changing streetscape on each site boundary. The associated public spaces are detailed to enhance the public realm on all four frontages and conceived as “gifts to the community”.

Much design thought has been given to enhancement of the patient experience in terms of improved accessibility, clear way finding, and the creation of a therapeutic environment. Courtyards have been increased in size to maximise the effect of natural light and views throughout, particularly within the major circulation and patient waiting areas. An integrated Healing Arts Strategy has been achieved, with architects and local artists working closely with the wider community to create a welcoming and stimulating healthcare environment which will act as a prime catalyst in the urban regeneration process in this historic part of Glasgow.

新维多利亚医院每年将治疗400,000患者，为他们提供综合诊疗服务，包括门诊、日间手术、复原、特别急救服务等，在南格拉斯哥引入了一个一站式医疗模式。项目具有一些原则：为本地社区门诊医疗提供更多的通道、阻止可选择护理效率的应急工作、使患者的诊疗过程流线化并提供一站式服务、达成医疗保健和技术的兼容性。

在这些要求下，新建筑与其独特的城市环境形成了积极的联系。为了减少周边交通噪声的影响、保证病人和工作人员的舒适感，医院被打造成了一个“密封盒子”，融入了完整的环境和舒适的降温设施。

建筑最大限度地保留了原有的成熟树木，立面与变换的街景相互映衬。相关的公共空间经过细部修整，提高了公共性能，被认为是“留给社区的礼物”。

设计师对提高病患的感官体验花费了大量的心思，改进了可达性、明确了指路系统，并且打造了益于健康的环境。庭院的扩张让自然采光和景观效果达到最佳，尤其是在主走廊和病患等候区。建筑师和当地艺术家一起打造了康复艺术策略，营造了热情洋溢的医疗环境。医院将成为格拉斯哥旧城区复兴的主要催化剂。

1. Urban context
2. Hospital in the park
3. Integration of art work into public realm

1. 医院所在的城市环境
2. 园区中的医院
3. 公共区域中的艺术作品

1

2

3

1. Outpatient crescent
2. Links with the past – historical continuity

1. 门诊半月造型
2. 与过去相连——历史的延续性

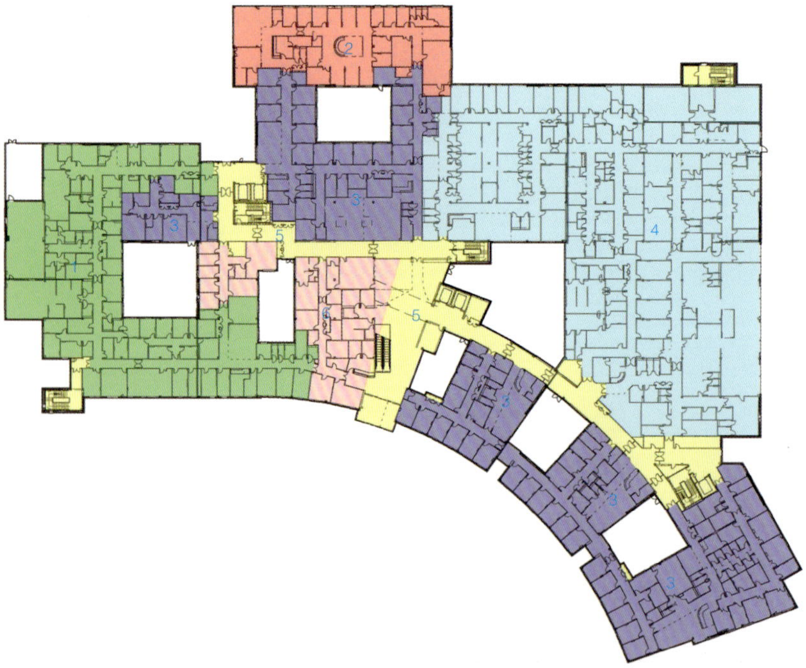

1. Rehab & day hospital
2. Minor injuries
3. Imaging & endoscopy diagnostics
4. Outpatients
5. Communication
6. Main entrance, support accommodation & medical records

1. 康复日间医院
2. 轻伤处理室
3. 成像和内窥镜检查
4. 门诊
5. 交流室
6. 主入口，辅助病房和病历记录室

2

1

1. Integration of artwork into building fabric
2. Hospital in the park
3. Maximum use of natural light
4. Atrium link bridge

1. 艺术品被结合到建筑之中
2. 园区中的医院
3. 最大限度使用自然采光
4. 中庭连接桥

1

2

1 Clinic waiting area with integration of healing arts
2. Public street view from atrium

1. 候诊区装饰着具有治愈功能的艺术品
2. 从中庭看公共街道

首尔圣玛丽医院 Seoul St. Mary's Hospital

Location: Seoul, South Korea **Completion date:** 2009 **Designer:** Samoo Architects & Engineers **Photographer:** Park Young-chae **Area:** 188,100 m^2

项目地点：韩国，首尔 完成时间：2009年 设计师：萨姆建筑工程公司 摄影师：朴杨在 面积：188,100平方米

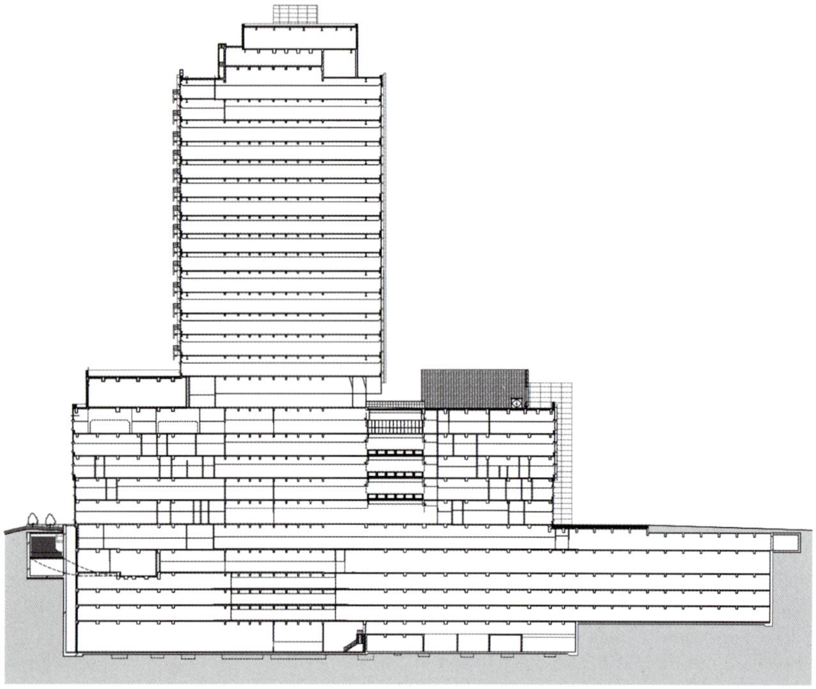

Built inside the Catholic Medical University, what Seoul St. Mary's Hospital needed in the midst of intensifying competition in the medical market and the surging demands from patients is a long-term development plan. This is the starting point of drawing up a plan for Seoul St. Mary's Hospital which is on its way to become not only the cornerstone of the Catholic Medical Science Complex but also the finest-class professional medical centre. Specifically, Seoul St. Mary's Hospital has been designed to be the multi-medical centre that combines the School of Medicine of the Catholic University of Korea and its former establishment Gangnam St. Mary's Hospital and offers the medical experts the most optimal research, education and medical functions while providing support to patients with pleasant and comfortable healing environment.

This is a 22-storey huge hospital building, amounted to 188,100 square metres. It also has six storeys below. It is the largest hospital in Korea. Compared with other catholic medical centres around the globe, its size, facilities and infrastructure are one of the best in the world.

A friendly and warm image is created through segmented masses and a careful selection of materials. While the terracotta panels in the lower floors bring a sense of warmth and stability, the aluminum panels and curtain walls infuse the upper floors with enterprising and forward-looking images. Most notably, the Hospital Street which is set between the central medical department and the outpatient department is not only regarded as the convenient way-finding to patients but also adds a surplus of natural lights to create the space that is brighter and more pleasant. The interlayer slabs facing this area have been applied with plywood of diverse sizes and thicknesses for 108 types of combinations, producing a sense of warmth that only wooden materials can bring.

首尔圣玛丽医院位于天主教医科大学内，在日渐激烈的医疗市场和病患日渐增长的需求中，医院必须有一个长期的发展规划。首尔圣玛丽医院的规划不仅是天主教医疗科学综合体的基石，还是一所顶级的专业医疗中心。作为一所多学科医疗中心，首尔圣玛丽医院合并了韩国天主教大学医学院和江南圣玛丽医院，为医疗专家提供最佳研究、教育和医疗设施，也为患者提供舒适愉悦的治疗环境。

医院高22层，总面积188,100平方米，地下还有6层空间，是韩国最大的医院。与其他国家的天主教医疗中心相比，它的规模、设备和基础设施都是世界顶尖的。

分段结构和材料的选择使医院看起来友好而温和。底层的赤陶板带来了暖意和稳重感，而上层的铝板和玻璃幕墙则充满了进取和超前的意味。值得注目的是，位于中心医疗部门和门诊部之间的医院街不仅为患者提供了简明便利的东路，而且为室内带来了充足的自然光，让空间看起来更明亮、更愉悦。拥有108种组合方式的尺寸、厚度不一的夹层板为这一区域带来了木材特有的温馨感。

1. A sculpture in front of the building
2. The entrance
3. The building is divided into two units
4. The use of segmental structure and selection of materials make the hospital friendly and warm

1. 建筑前方的雕塑
2. 医院入口
3. 建筑分为纵横的两大模块
4. 分段结构和材料的选择使医院看起来友好而温和

Award:

(Royal Institute of Chartered Surveyors) RICS Scotland 2010 Awards – Sustainability Project of the Year / Overall Project of the Year

Scottish Design Award 2010 – Commercial Project category

获奖情况：

2010年皇家特许测量师学会苏格兰测量师学会——年度可持续项目奖/年度项目

2010年苏格兰设计奖－商业项目类

1

2

3

4

1

1. The hospital has a sense of modern design
2. The interior atmosphere is comfortable and graceful

1. 医院设计现代感十足
2. 内部氛围温馨高雅

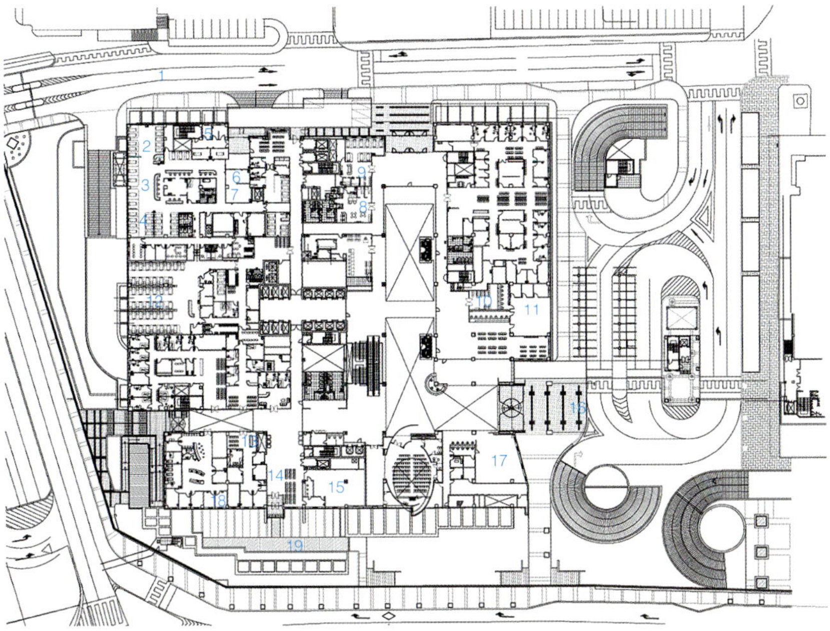

1. Emergency centre
2. Survey patient
3. Serious case
4. Acute patient
5. Emergency resource and interview room
6. CPR room
7. External wound revival room
8. Examination room before hospitalisation
9. Injection room
10. Formalities for hospitalisation and leaving hospital
11. Administration
12. Dialysis room
13. Outpatient surgery clinic
14. Waiting of pharmacy
15. Preparation of medicine for outpatient
16. Lobby
17. Café
18. Medical office
19. Deck

1. 急救中心
2. 手术病房
3. 重症室
4. 重症病房
5. 紧急资源和接待室
6. 心肺复苏室
7. 外伤复原室
8. 住院前诊断室
9. 注射室
10. 住离院手续办理处
11. 行政区
12. 透析室
13. 门诊手术室
14. 药房等候区
15. 门诊医疗准备区
16. 大堂
17. 咖啡厅
18. 医疗办公室
19. 平台

1 A simple and lively reception
2. Resting area
3. The Catholic room

1. 简洁明快的接待台
2. 休息区
3. 天主教教堂

2

3

圣文森特医疗中心 St. Vincent's Medical Centre

Location: Bridgeport, USA **Completion date:** 2010 **Designer:** Perkins Eastman **Photographer:** Chris Cooper, Sarah Mechling **Area:** 11,612 m^2

项目地点：美国，布里奇波特 完成时间：2010年 设计师：珀金斯·伊斯曼 摄影师：克里斯·库伯、沙拉·马奇林 面积：11,612平方米

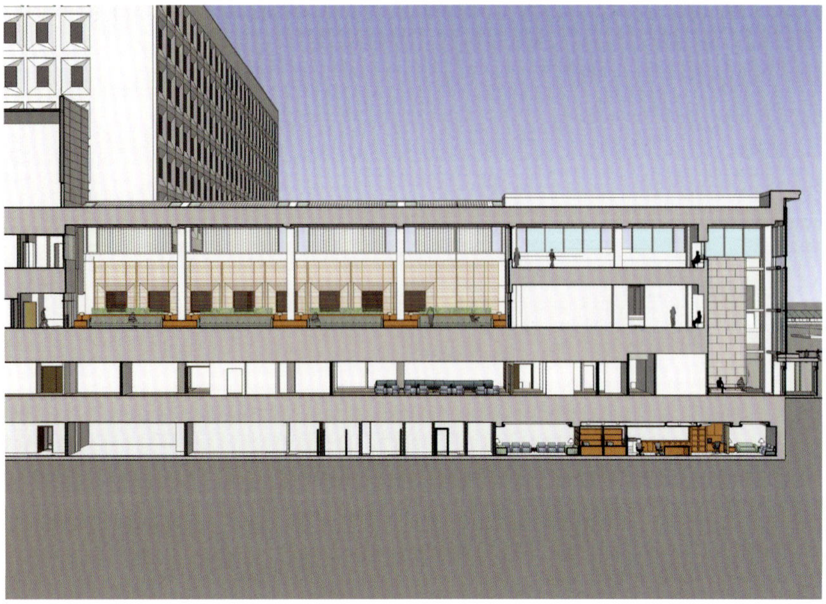

As the centrepiece of a facilities master plan implementation for St. Vincent's Medical Centre in Bridgeport, this new four-storey facility houses a consolidated, centralised cancer centre and an expanded emergency department. It is the most significant expansion project in the Medical Centre's 106-year history – positioning St. Vincent's Medical Centre to support the increasing population of the community for years to come.

St. Vincent's Medical Centre (SVMC) has been an important presence in Bridgeport for more than 100 years. Over time, the institution expanded and replaced existing buildings on the original site. With the inception of the facilities master planning effort, SVMC considered building off-site, into the suburbs where space was more available. However, during the planning process, it became apparent that their commitment to the city coupled by strong desires to cluster the acute care block and "centres of excellence" all at the same location required expansion within the existing urban site.

Accordingly, the master plan presented a solution that accommodated multiple needs. Four programme areas in need of growth were identified: emergency medicine, cancer care, parking, and inpatient beds. Through careful placement of the new 600-space parking garage on the southwest corner of the site, a campus courtyard was created with a formal landscaped allee of oak trees, linking the existing staff parking garage to the main hospital entrance while creating a ceremonial gateway to Main Street. Also part of the plan, the main floor of the Medical Centre was reorganised, creating an extended lobby concourse that provides easy access to important adjacent centres of excellence: outpatient surgery, imaging, and cardiology while also providing an important link to the new cancer centre. The master plan also incorporated the future vertical expansion of the hospital on top of a two-storey addition, permitting the construction of a future bed tower. When realised, this addition will provide much needed single-bed rooms while allowing for the renovation of the existing inpatient facilities. The present construction incorporated new columns and footings to support the future vertical addition.

作为布里奇波特圣文森特医疗中心总体规划的核心部分，这座四层楼高的医疗机构包含一个完整的癌症中心和一个扩建的急诊部。它是医疗中心106年历史上最重要的扩建项目，让圣文森特医疗中心得以应对未来社区人口的不断增长。

近百年来，圣文森特医疗中心一直在布里奇波特起到了重要的作用。多年以来，中心一直在进行扩建，在原有场地上进行建筑的拆迁与重建。在总体规划中，圣文森特医疗中心考虑将新设施建在空间充裕的城郊。但是在规划阶段，事实证明急症护理楼和“卓越中心”必须设在同一地点，需要在现有的市区场地里进行扩建。

因此，项目的总体规划为多方面需求呈现了解决方案，共有四个明确的区域：急救区、癌症治疗区、停车场和住院病房。600个停车位的车库位于场地的西南角，两侧种满橡树的景观大道形成了一个庭院，连接了原有的员工车库和医院的主入口，也打造了通往梅因大街的正式通道。医疗中心的主楼经过了重新布局，打造了一个中央大厅，提供了通往各个部门的方便通道，如门诊手术室、成像区、心脏科和癌症中心。总体规划扩建了医院的垂直建构，在上层增加两层楼，以建造新的病房。实现之后，这个扩展空间将提供更多的独立病房，还能保证原有住院设施的翻新。目前的建造工程包含支撑未来垂直扩建的新支柱和地基。

1. The white building is clean and lively
2. A night view of the medical centre
3. The entrance with fan-shaped roof

1. 白色建筑简洁明快
2. 夜色中的医疗中心
3. 扇形顶棚的入口

1

2

EMERGENCY & TRAUMA CARE
CENTER FOR CANCER CARE
STOP

1

2

3

4

1. The façade of the medical centre
2. The courtyard landscape in wood
3. The spacious resting area
3. The entrance

1. 医疗中心正面投影
2. 木质结构的庭院景观
3. 开阔的休息区
4. 入口

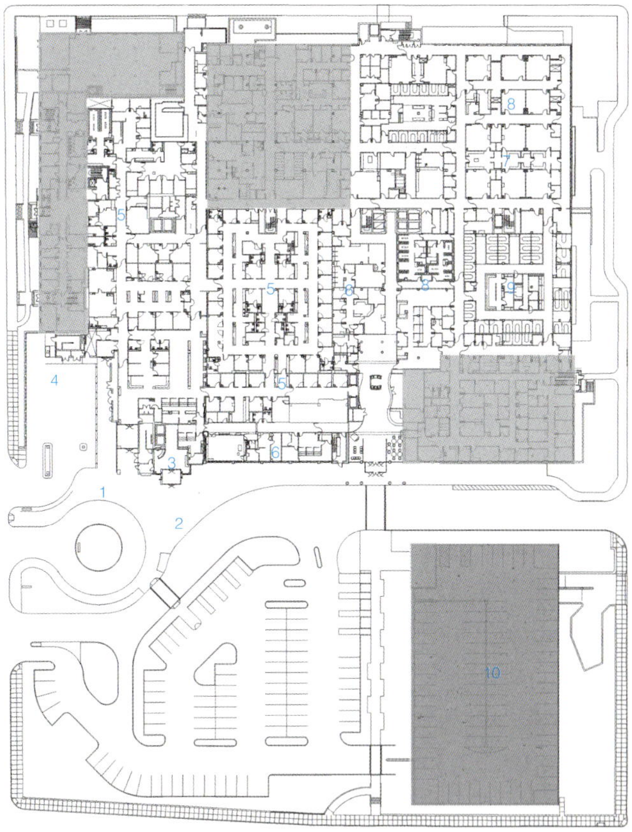

1. Emergency department drop off
2. Cancer centre entrance drop off
3. Cancer centre entrance
4. Ambulance bays
5. Emergency department
6. Common and support
7. Cardiology
8. Ambulatory
9. Cardio/Amb Shared
10. Garage (earlier phase)

1. 急诊科下车口
2. 癌症中心下车口
3. 癌症中心入口
4. 救护车停车处
5. 急诊科
6. 公共和辅助设施
7. 心脏科
8. 回廊
9. 有氧/电磁室
10. 车库（早期）

ton B. Hollander

1. The open and spacious hall
2. The interior design has a home-like atmosphere
3. Waiting area

1. 宽敞开阔的中央大厅
2. 室内设计居家气息浓厚
3. 候诊区

5

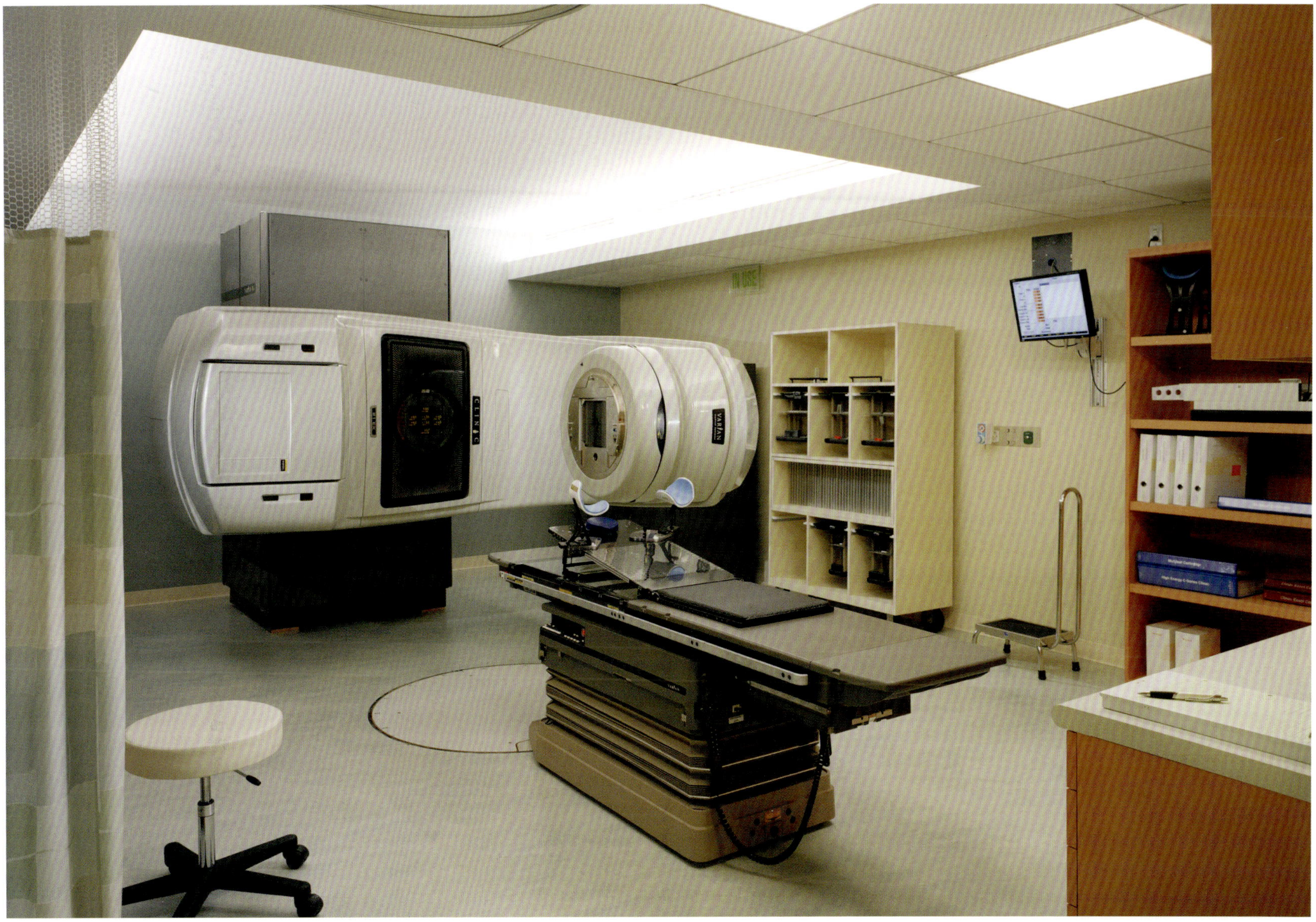

1. Resting area
2. Wood is the main material in interior design
3. Medical area
4. Corridor

1. 休息区
2. 室内设计主要采用木制材料
3. 医疗区
4. 走廊

竹川医院 Takekawa Hospital

Location: Tokyo, Japan **Completion date:** 2007 **Designer:** Kidosaki Architects Studio **Photographer:** 45g Photography-Junji Kojima **Area:** 4,056 m²

项目地点：日本，东京 完成时间：2007年 设计师：城户崎建筑研究室 摄影师：小岛纯司 面积：4,056平方米

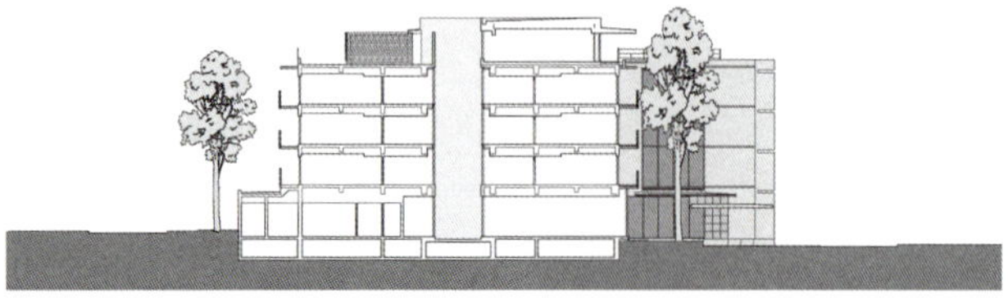

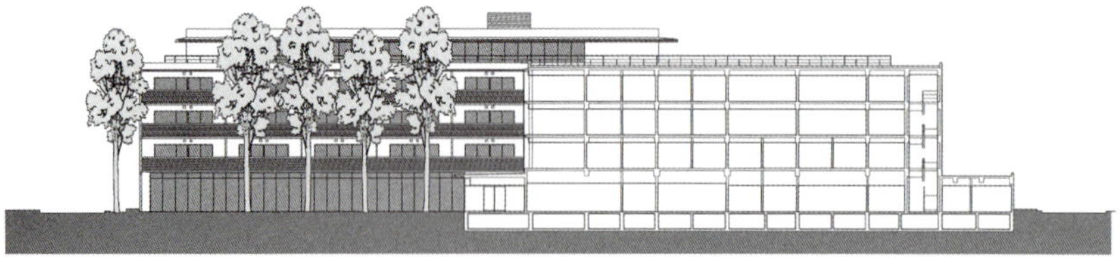

The site is a corner enclosed by the very few rich green of big Zelkova trees in Tokyo. Takekawa Hospital/Care Centre Keyaki were planed as a base of the regional medical institution for elderly and for the welfare, complementing mutually each function of the medical treatment hospital with 151 beds and Care Centre with 28 beds.

To proceed design study, the designers made it a rule to preserve the Zelkova trees which have coexisted with the site for over 400 years and to coexist with the architecture. To match the scenery, the designers make the height of the architecture lower than that of the Zelkova trees; besides, they setback the 4th floor, and put it close to the trees. Therefore, the architecture shows an appearance of being incorporated in the Zelkova forest.

For landscape, the designers considered contribution for a regional community. The aim was to make harmonious surroundings with private sidewalk circling the Zelkova trees, and a well was installed for preparation for future disasters, if any.

Takekawa Hospital and Care Centre Keyaki both have full-height glasses at the end of the circular tour corridor to view the Zelkova trees and the blue sky, giving open and light impression into the building. The light garden is not only for lighting; it has the function of an open-air shaft which contributes to improve condition of the surroundings.

Generally people feel the hospital room is small; therefore, the designers use structure systems which unify the column and the structure wall to highlight the expanse of space. Each room's balcony is possible for barrier-free access. When the window is open, the balcony can be used as amenity space with daylight streaming through the leaves of the Zelkova, not only for extension of room. Each room is equipped with a toilet and a washstand to urge the patient to become independent.

The 4th floor's rehabilitation area (a main function of Takekawa Hospital) encircled by glass windows has the deep eaves. Moreover, this area can feel the expanse of space by being unified with the terrace which is possible for barrier-free access. The terrace commands a panoramic view of the newly-developed city centre of Shinjyuku. The architecture completed which has both dignity and fine quality is a good example of low-cost construction by uncompromising control of detail with careful deliberation and sophistication, using ready-made materials in various places effectively.

竹川医院/榉树护理中心位于东京，被一片茂密的榉树所包围，专门为老年人和福利人群提供医疗服务，医院有151个床位，而护理中心有28个。

在设计过程中，设计师保存了与场地共存400年以上的榉树，使其与建筑和平共处。为了与景色相配，建筑的高度要低于榉树，并且五楼向内缩入，与树木接近。这样一来，建筑看起来就像与榉树林合为一体。

设计师试图让景观设计对社区做出贡献。其目的是环绕着榉树打造人行道，并且为预防灾难降临打一口井。

竹川医院和榉树护理中心在环形走廊的尽头都有落地窗，可以看到榉树和蓝天的景色，让建筑看起来开放而清新。日光花园不仅能够提供自然采光，还起到了通风作用，为周边环境的设计做出了贡献。

人们总是认为医院的房间很狭小，因此设计师运用统一的支柱和结构墙系统来让空间显得开阔。打开窗户，阳台可以作为休闲空间，透过榉树树叶缝隙照射过来的阳光让人心情愉悦。每间病房都配有马桶和盥洗盆，以便让病人学会独立。

五楼的康复区（竹川医院的主要功能区）由玻璃窗围绕，顶上是深深的屋檐。平台的运用延伸了康复区，保证了无障碍出入。平台可以俯瞰重建的新宿区全景。这座低成本建筑看起来庄重而品质优良，在细节上深思熟虑，利用现成的材料让各个区域具有较高的效率。

1. A bird's-eye view
2. The architecture coexists peacefully with the Zelkova trees
3. Through the roof, the Zelkova trees merge into the architecture
4. A side view of the building

1. 俯瞰图
2. 建筑与榉树和平共处
3. 榉树穿过顶棚与建筑融为一体
4. 建筑侧立面

1

2

3

4

1

2

3

4

1. Grey and white are the main colours of the building
2. Details of the windows
3. The 4th floor's rehabilitation area is encircled by glass windows
4. The terrace on the roof for rest

1. 建筑主体由灰白两色构成
2. 玻璃窗细节
3. 五楼的康复区由玻璃窗围绕
4. 顶层平台的休息区

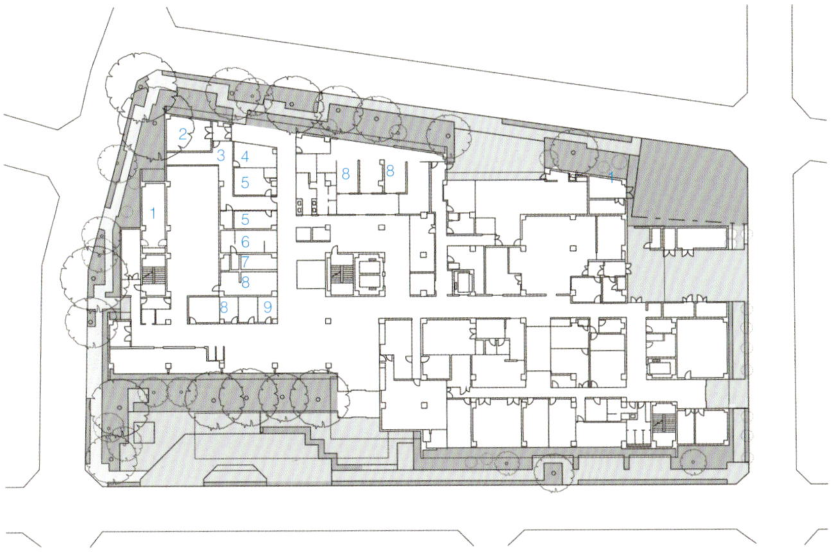

1. Machinery room
2. MR
3. Operating room
4. CT
5. X rays
6. Endoscope
7. Ultrasonography
8. Consultation room
9. Barber

1. 机械室
2. 核磁共振室
3. 手术室
4. 断层扫描室
5. X射线室
6. 内窥镜室
7. 超声波室
8. 诊察室
9. 理发室

1

2

1. The building is surrounded by trees
2. The corridor
3. The resting area has a good lighting

1. 树木环绕的医院主体
2. 走廊
3. 休息区采光良好

1

2

3

1. Medical area
2. The pathway
3. The simple and generous patient room

1. 医疗区
2. 通道
3. 病房设计简单大方

挪威放射医院 The Norwegian Radium Hospital - RADIATION THERAPY

Location: Oslo, Norway **Completion date:** 2006 **Designer:** Henning Larsen Architects **Photographer:** Thorbjorn Hansen **Area:** 8,500 m^2

项目地点：挪威，奥斯陆 完成时间：2006年 设计师：亨宁·拉尔森建筑事务所 摄影师：托尔比乔恩·汉森 面积：8,500平方米

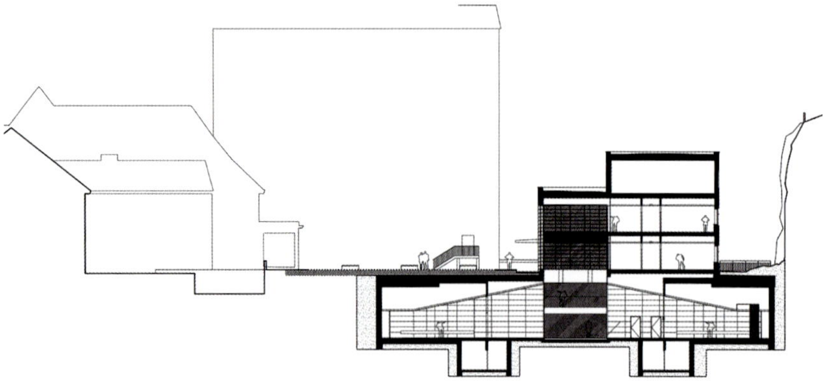

The Radium Hospital is the central cancer treatment and research institution in Norway and is located in a hilly and densely built-up area in Ullern, Oslo.

The radiation treatment department is located on the highest and most hilly part of the site. It comprises five new radiotherapy facilities, examination functions partly below terrain as well as laboratory, teaching and office facilities above terrain.

The oldest buildings of the hospital, the new extension and the tall trees of the nearby Mærradal provide the framework of a new local arrival's area from where patients have direct access to the facilities of the building. The existing main entrance of the hospital has level free access to the four-storey high and bright foyer.

A high south-facing glass façade ensures a rich inflow of daylight into all building functions. The flora and stream of the Mærradal is visible through the large west-facing windows in the resting area of the foyer and from the examination and conversation rooms of the lower west wing.

The light timbered walls and sloping ceilings of the entrance areas and treatment rooms reflect the daylight far into the rooms. The angled shape of the entrance area makes it possible to avoid heavy metal doors and instead use tall glass doors as the only separating element during treatments.

放射医院是挪威的中央癌症治疗和研究机构，位于奥斯陆的山地楼宇密集区。

放射疗法科位于当地最高的山地上，拥有5台全新放射机、半地下检查室和地面上的实验室、教室、办公室。

旧医院大楼、新扩建部分和高大的树木环绕出一个新的到达广场，病人可以由此直接到达各个科室。原有的主入口直接通往宽敞明亮的大厅。

朝南的玻璃幕墙保证了建筑充足的自然采光。透过大厅休息区西面巨大的玻璃窗和西翼的检查室和会谈室可以看见山上的植物和泉水。

入口区域和诊疗室浅色的木墙和斜面天花板将日光引入房间里。入口区的角度没有使用厚重的金属门，而采用了高大的玻璃门，这是治疗中唯一的分隔元素。

1. The glass curtain wall is transparent and bright
2. A side view of the building
3. The glass façade ensures a rich inflow of daylight

1. 玻璃幕墙通透明亮
2. 建筑侧面
3. 巨大的玻璃保证了良好的自然采光

1

2

3

1

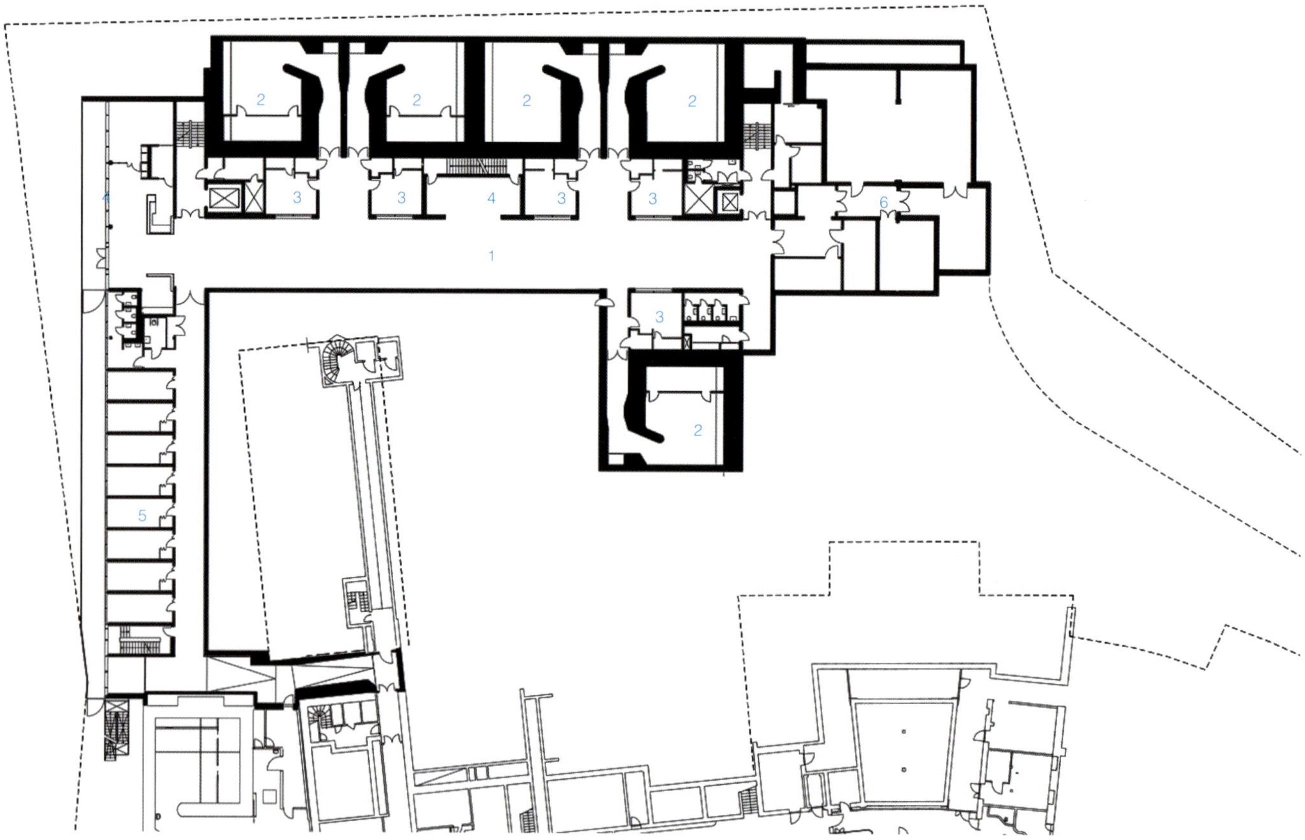

1. Lysholm
2. Radio therapy
3. Control
4. Waiting area
5. Examination rooms
6. Technique

1. 评估室
2. 放射疗法室
3. 控制室
4. 等候区
5. 检查室
6. 技术室

2

1. The hospital and environment are combined together
2. The open and spacious hall
3. Internal pathway

1. 医院与环境景观紧密的结合在一起
2. 大厅宽敞开阔
3. 内部通道

3

1. The open corridor leads to the different functional areas
2. The upper floors use glass structure

1. 开阔的走廊通往各个功能区域
2. 建筑上部采用了玻璃构造

2

1

2

1. Resting area
2. Conference room has a good dayligt
3. Details of the conference room

1. 休息区
2. 会议区采光良好
3. 会议区细部

3

费索尔医院 Vesoul Hospital

Location: Vesoul, France **Completion date:** 2009 **Designer:** Groupe-6, with INGEBAT and Jacobs France **Photographer:** Luc Boegly, Eymard Duvernay **Area:** 52,000 m^2

项目地点：法国，费索尔 完成时间：2009年 设计师：6组设计、INGEBAT公司、雅各布斯·弗朗斯 摄影师：卢克·伯格利、爱玛德·杜沃内 面积：52,000平方米

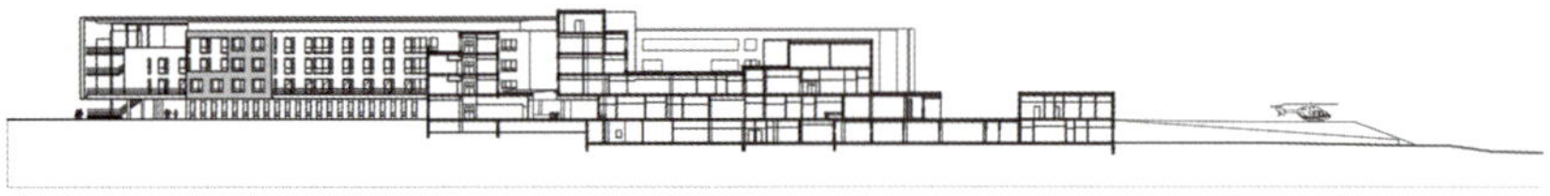

The Vesoul Hospital (CHIHS) merges the services of the three main sites of Vesoul, Lure and Luxeuil, within a new building located West of Vesoul. The building offers a triple face: The serenity and domesticity of the wards turned towards the city and the emblematic hill of the city to the south. Opened onto the green park, this façade is freed from any traffic nuisance. The technical platform has an industrial look and the logistics are with red concrete blocks and metallic boxes. The tin roofed curve of the offices follows the ring road and protects from the sound nuisances

Emphasis has been given to the interior design and signage work, offering the staff and patients a surprisingly new and comfortable environment. The classic white universe of the hospital is made less traumatic, animated here by a rich polychromy and decorated with contemporary furniture closer to the "real world". 80% of the bedrooms are individual rooms, each one of them uniquely defined by a specific animal, like a protecting totem.

Beyond the plastic quality of the façades and the interior atmosphere, the new organisation is easy to understand. Clarity of the structure, transparency, vertical and horizontal paths: while banishing uniformity, everything works naturally towards the creation of a simple and pleasant working and living space. On three levels, the building keeps a human scale and offers an easy proximity between services with a shortening of distances down to 25 metres maximum.

费索尔医院位于费索尔西部，为费索尔、鲁尔和鲁西尔三个地区提供医疗服务。建筑分为三个层面。朝向城市和城市南部标志性小山的病房拥有宁静感和家居感，朝向公园，远离交通污染。技术平台采用了工业化造型，而后勤部是红色混凝土砖块结构和金属结构。办公楼的锡屋顶曲线与环形公路相对应，并且能阻隔噪声污染。

设计重点在于室内设计和引导标示设计，为工作人员和患者提供一个全新而舒适的环境。室内弱化了经典的白色色调，通过丰富的色彩和现代化的家具让空间更加生动，贴近现实生活。80%的病房是独立病房，每间都有一个独立的动物主题，就像守护图腾一样。

除了外墙和室内氛围的品质之外，新的组织结构也简单易懂。简明的结构、透明感、垂直和水平路径虽然并不一致，却自然地打造了一个简单而舒适的工作和生活空间。建筑的三层楼都保持了人性尺度，距离服务区的最远距离缩减到25米。

1. A long-range perspective
2. Wing-shaped inpatient ward
3. The logistics department with red concrete blocks and metallic boxes

1. 建筑远景
2. 翼状住院病房
3. 红色混凝土砖块结构和金属结构组成的后勤部

1

2

3

1

2

1. The interior design is fluid and bright
2. The corridor is defined by specific animals

1. 室内设计明快流畅
2. 走廊墙体印有独立动物主题

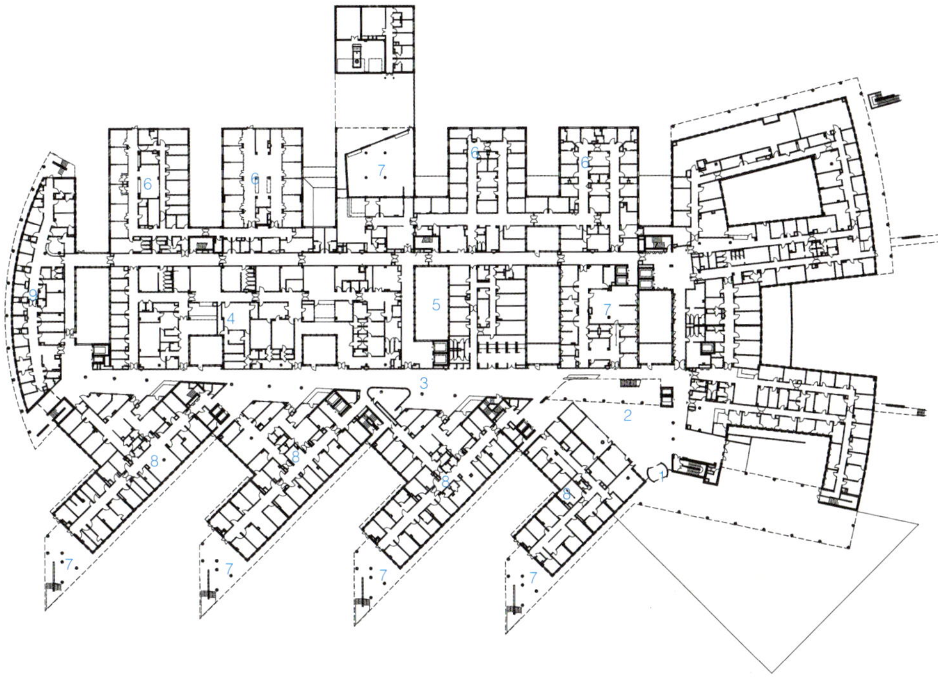

1. Main entrance
2. Entrance hall
3. Corridor
4. Operating rooms
5. ICU
6. Consulting rooms
7. Waiting area
8. Sickrooms
9. Nurse stations

1. 主入口
2. 入口大厅
3. 走廊
4. 手术室
5. 特护病房
6. 诊疗室
7. 等候区
8. 病房
9. 护理台

1. Green stairs and white walls contrast each other
2. The lively designed central hall

1. 绿色的楼梯与白色墙面形成鲜明对比
2. 设计活泼的中央大厅

Bâtiment Air
munal de Haute-Saône
Bâtiment Eau
Site de Vesoul
Accueil
5
3

2

1

Bâtiment Terre
Bâtiment
Centre hospitalier intercommunal de Haute-Saône
2

3

1. The transparent materials add brightness to the interior
2. The use of wood gives the patients a warm feeling
3. A general view of the atrium

1. 楼梯采用透明材料增添了室内的通透感
2. 木材的使用带给患者亲切感
3. 中央大厅纵览

提顿放射科——麦迪逊成像中心

Teton Radiology – Madison Imaging Centre

Location: Rexburg, USA **Completion date:** 2009 **Designer:** Ward+Blake Architects **Photographer:** Roger Wade **Area:** 1,064 m^2

项目地点：美国，雷克斯堡 完成时间：2009年 设计师：瓦德+布莱克建筑事务所 摄影师：罗格・韦德 面积：1,064平方米

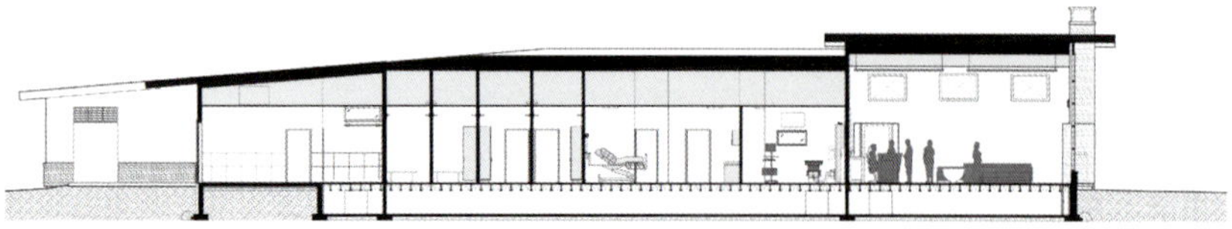

Teton Radiology – Madison Imaging Centre is located in a business park setting on the north end of Rexburg, Idaho. Clients' initial desire was for a facility at the leading edge of the profession with a matching medical practice mission combining the following: patient-centred focus making the radiology experience comfortable and private for primarily female clientele; cutting-edge medical technology and procedures; sustainability regarding environmental sensitivity and in the facility's ability to evolve with business and technological changes; and sense of locale placing the structure clearly within the small-town Idaho community with rural roots.

Architects selected the "spa" design model after looking to the hospitality industry for approaches consciously associated with wellness. A natural-light flooded "oversize" lobby incorporating low-maintenance yet sophisticated materials like natural stone and wood invites comfort with cues from familiar contemporary aesthetic. privacy theme underlies the plan that flows clearly from the lobby to the second waiting room for women with private bathroom and changing rooms corresponding with the mammography and ultrasound rooms. Separate porte-cochere exit allows patients to leave without returning to the lobby, reflecting sensitivity to potential physical and emotional effects of treatment/diagnoses.

Facility breaks the "dark box" look of many imaging centres with eco-friendly emphasis on natural light invited via exterior glazing and interior skylights. Clerestory windows throughout the corridor bring light from exterior windows, and "Solar Tube" skylights brighten interior spaces. Two-stage corridor sensors adjust lighting according to natural levels. Materials throughout were selected for durability and reduced maintenance including recycled content in vinyl flooring and in easy-to-repair carpet tiles. The project is super-insulated to reduce energy load, and an efficient Daikin air to air heat exchanger controls the forced air mechanical system temperature. High-efficiency Solarban 70 glazing reduces heat gain, energy loss and glare; accomodates overhanging eaves a further heat gain reduction. Design flexibility accommodates for a future nuclear medicine room and hot lab in the current employee lunch room.

提顿放射科——麦迪逊成像中心位于爱达荷州雷克斯堡北部的商业园中。委托人的目的是打造一个顶尖的医疗设施并提供相匹配的顶级医疗服务，项目将结合以下目标：以患者为中心，提供舒适而私密的（尤其是为女性患者保密）放射性治疗服务；提供前沿医疗技术和程式；打造环境敏感度和设施围绕商业和技术转变的灵活性；打造简明清晰的建筑结构，使其与爱达荷州的小镇格局相适应。

建筑师选择水疗馆作为设计模型，试图将服务业于医疗业联系在一起。自然光洒在大尺寸的大厅中，壁炉由耐用而精致的天然石材和木材制成，体现了现代美学。私密主题在女士第二等候室、多功能诊断室的私人洗手间和放射科、诊断科的更衣室中得以体现。独立出入口让患者可以不用重返大厅就走出去，保护了患者在治疗和诊断后敏感的情绪。

成像中心避免了传统的"暗箱"形象，透过玻璃幕墙和室内天窗洒进的自然光让它更加环保。走廊上的纵向天窗带来了阳光，而"日光管"天窗则点亮了室内空间。两级走廊根据自然条件调整照明。全部的材料都选择了耐用而易于保养的，如回收乙烯基地板和易修复块式地毯。项目具有高绝缘性，减少了能量负载，高效的大金空调系统控制了室内的温度。高效的隔热玻璃减少了热辐射、能量流失和刺眼的光线。飞檐也对减少热辐射做出了贡献。根据需要，可以在现有的员工午餐室中灵活地建造核医学实验室和热实验室。

1. The façade of the building 1. 建筑外立面
2.The entrance 2. 入口处
3. The colour palette is harmonious 3. 建筑颜色搭配融洽

1

Teton Radiology
Madison
advanced medical imaging
2

Teton Radiology
Madison
advanced medical imaging
3

1

1. The fireplace incorporates sophisticated natural stone
2. The comfortable resting area with abundant natural light

1. 壁炉由耐用而精致的天然石材制成
2. 舒适的休息区采光良好

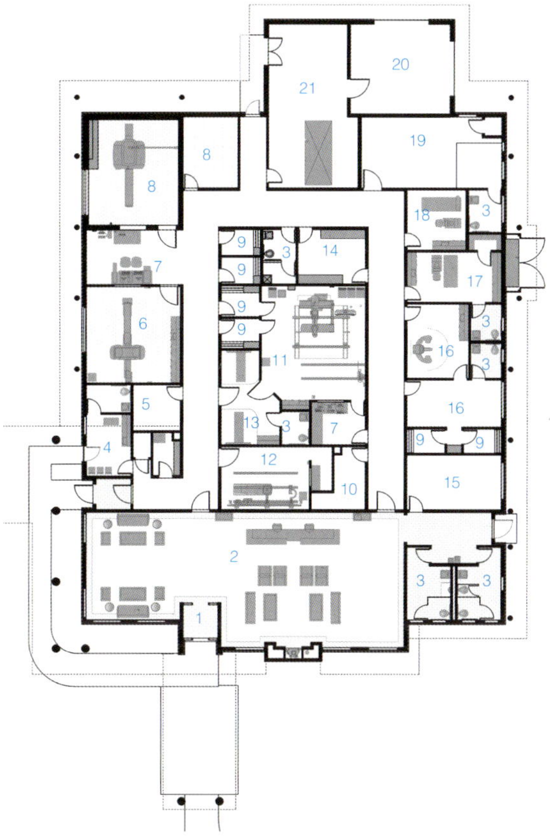

1. Entry
2. Lobby
3. Restroom
4. Lab
5. Computer
6. CT room
7. Control
8. MRI
9. Dressing room
10. Electrical
11. Multi-diagnostics
12. X-ray
13. Recovery
14. Reading room
15. Consulting
16. Mammography
17. Ultrasound
18. Bone density
19. Break room
20. Refrigeration room
21. Mechanical

1. 入口
2. 大堂
3. 洗手间
4. 实验室
5. 电脑室
6. 断层扫描室
7. 控制室
8. 核磁共振室
9. 更衣室
10. 电机室
11. 多层诊断室
12. X射线室
13. 恢复室
14. 阅读室
15. 诊疗室
16. 乳房X射线室
17. 超声波室
18. 骨密度测试室
19. 休息室
20. 冷藏室
21. 机械室

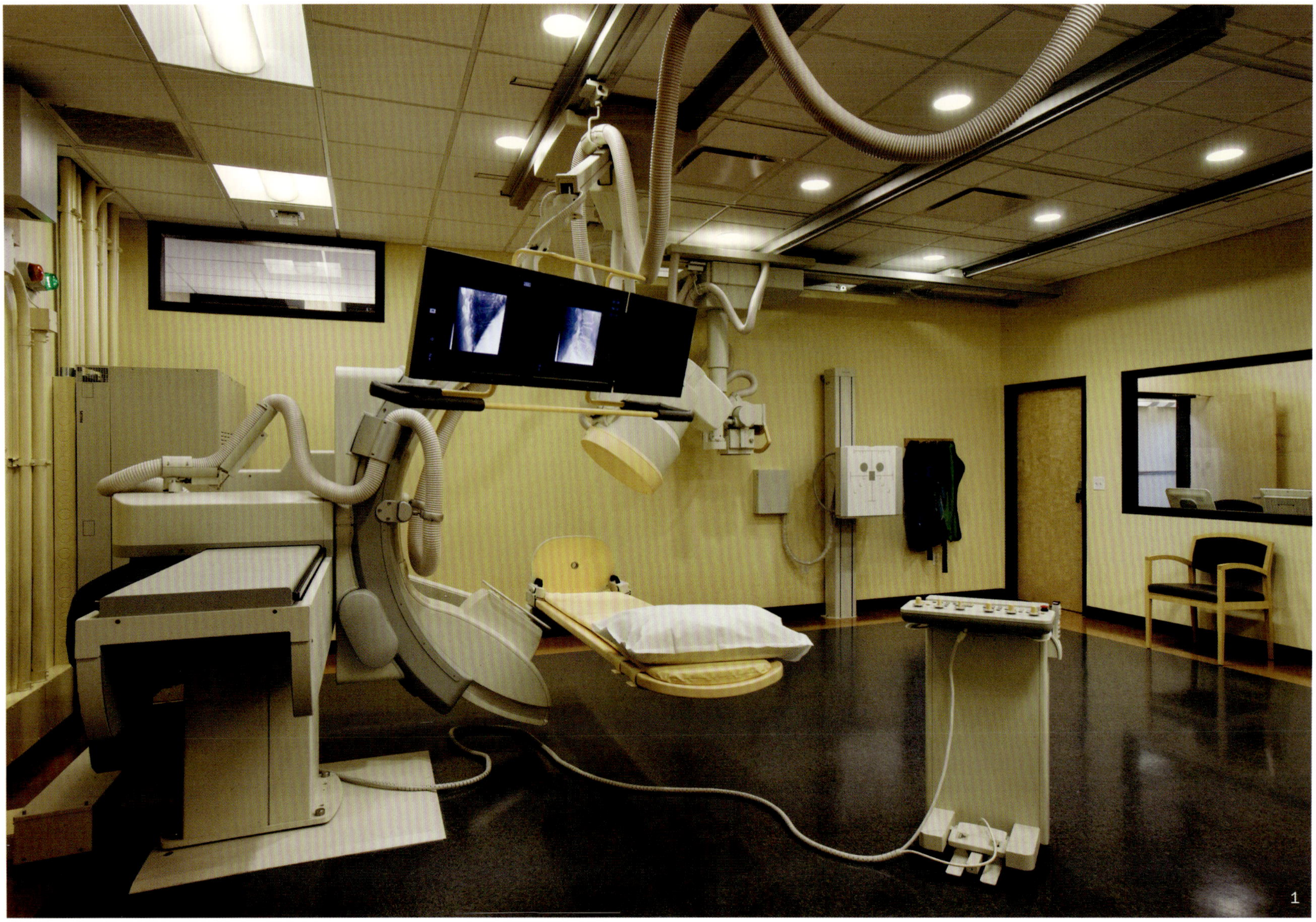

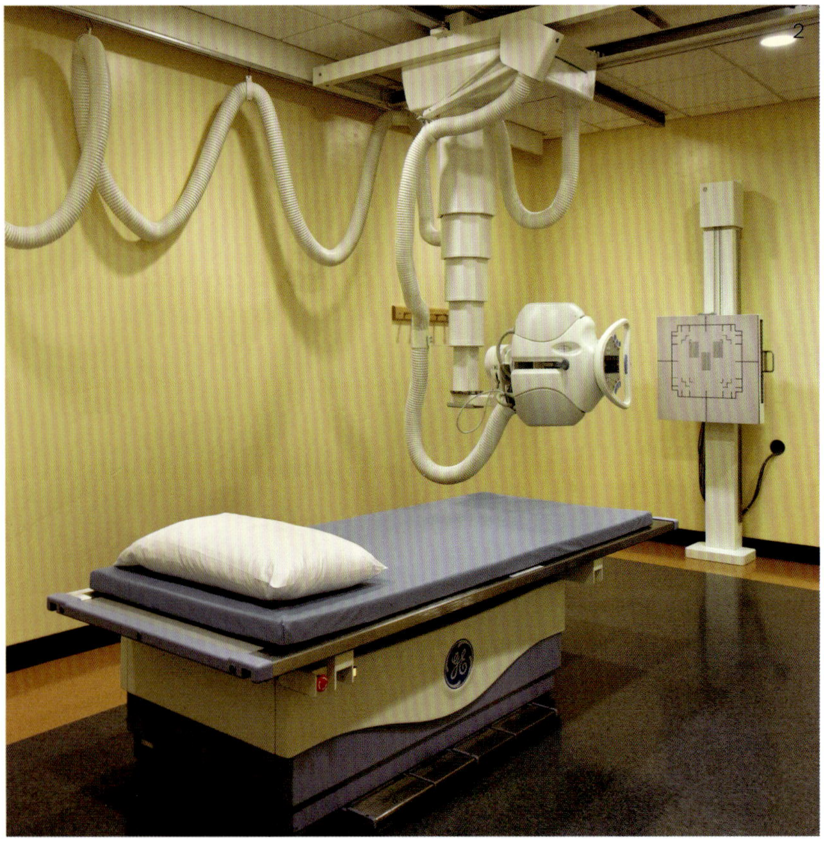

1. The complicated medical equipment (Multi-diagnostics room)
2. Medical area (X-ray room)
3. MRI room
4. Natural light enters the hallway via "Solar Tubes"

1. 复杂的医疗设施（多样诊断室）
2. 医疗功能区（X光室）
3. 核磁共振室
4. 自然光通过“太阳能管道”进入门厅

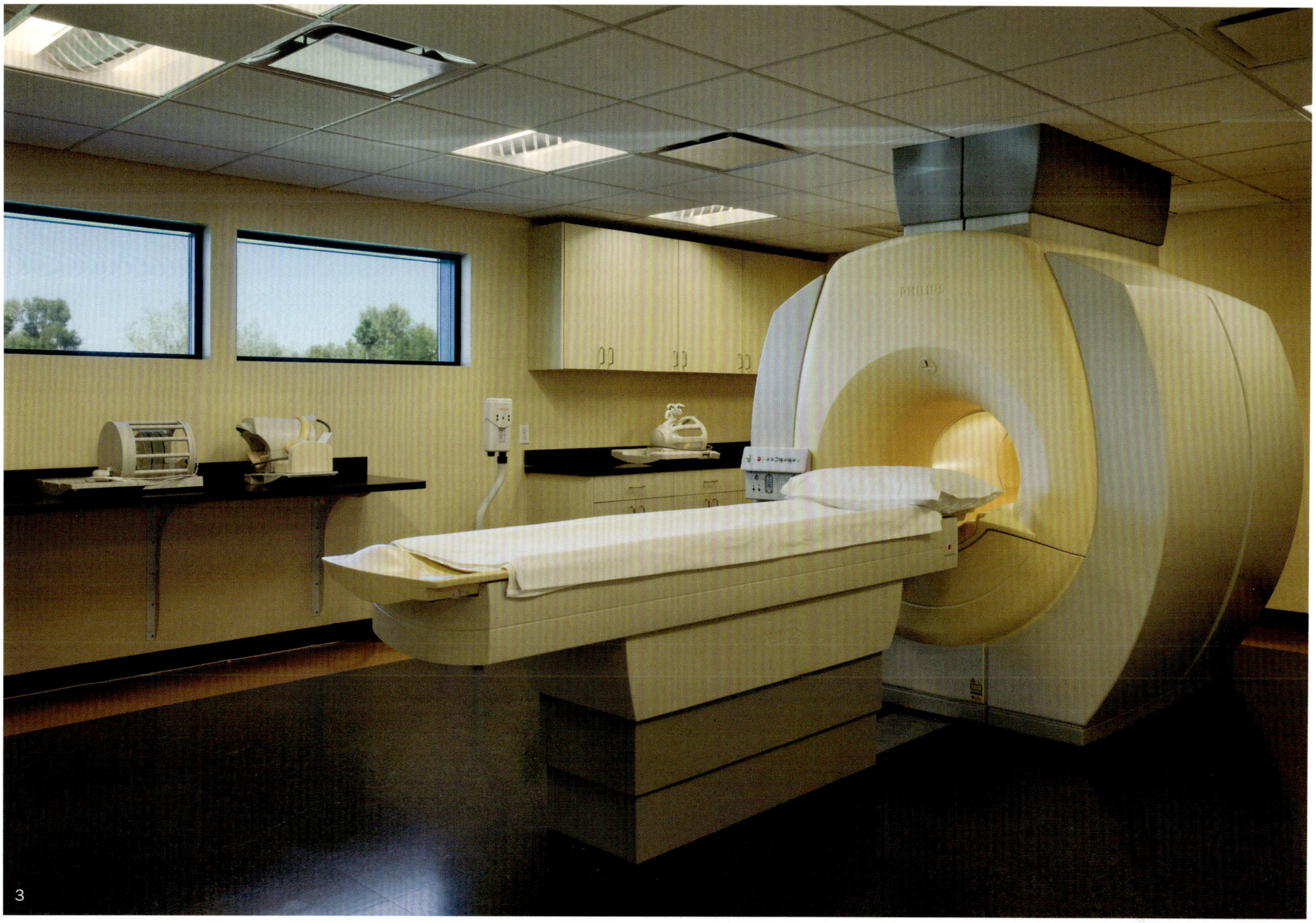

3

4

阿西巴代姆亚达那医院 Acibadem Adana Hospital

Location: Adana, Turkey **Completion date:** 2008 **Designer:** Acibadem Project Management **Photographer:** Cemal Emden **Area:** 20,000 m²

项目地点：土耳其，亚达那 完成时间：2008年 设计师：阿西巴代姆项目管理 摄影师：瑟梅尔·埃姆登 面积：20,000平方米

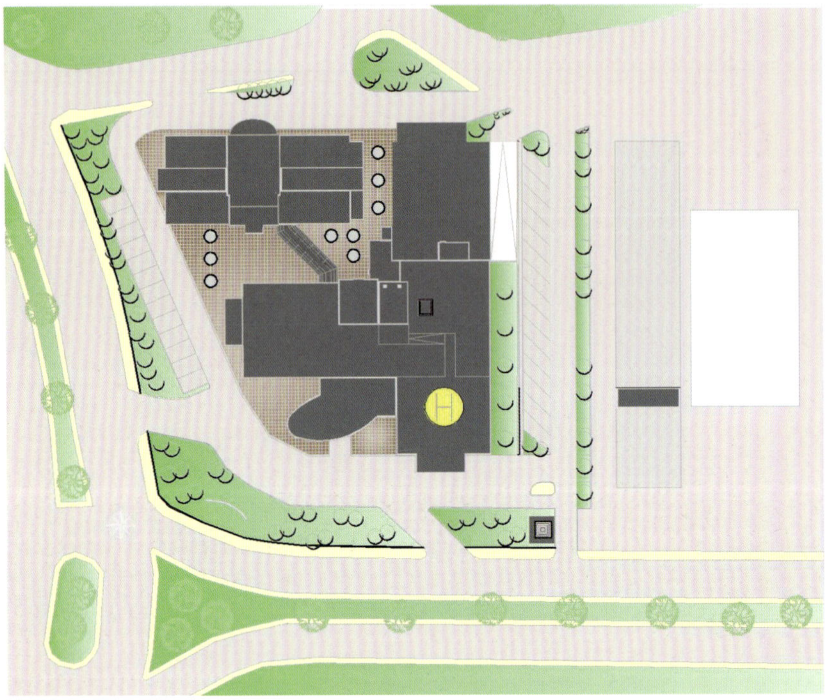

The 20,000-square-metre building located in Adana at the section of the city under renovation process was planned with 10 storey, and has been completed on 5,000-square-metre land. The hospital includes 22 clinics, 55 exam rooms, 120 beds, 26 intensive care rooms and 6 operating theatres.

The outer skin of the building was formed with structural glass surface cladding walls referring to the renovation process of the city, and supported with sunshades. The block including the emergency, main entrance and logistic entries were opened to the same street with the aim of preventing negative effect on the traffic. The interior equipments of the building were designed by association as both medical and interior decoration category with metropolitan hospital criteria, highly superior than the regional criteria. A big sluice at the entrance, and its extension, the lobby and its arrangement is inviting as well as planned to relieve the negative effects of the hospital restlessness on the patients. General natural materials were preferred, and the other industrial products used are within the upper segments in their categories. The wooden patterned laminate coverings, marbles, PVC flooring materials were applied by planning with the interpretation of consistent colours and patterns. In illumination, generally the lightings were applied putting the colours, objects and necessary functions in the foreground rather than the ordinariness of sharp and uncontrolled lightings washing the walls. Other planning items were designed and applied in line with the corporate approaches of Acibadem Healthcare Group.

建筑总面积20,000平方米，位于亚达那市的改造区，共有10层，占地面积5,000平方米。医院共有22间诊疗室、55间检查室、120个床位、26个特护病房和6间手术室。

建筑的表皮由结构玻璃墙组成，反映了城市的改造过程，并且有遮阳板支持。建筑包含紧急入口、主入口和后勤入口，全部朝向同一街道，以避免对交通产生负面影响。建筑的内部设施既考虑到了医疗功能又包含了都市医院装饰特点，远远高出当地的标准。入口的大水池和大厅的布置都十分吸引人，减轻了患者在医院里的紧张心情。

设计偏爱普通的自然材料，其他工业产品则选择了同类中的上乘。木纹薄板、大理石、聚氯乙烯地板材料采用了一致的色彩和图案。在照明方面，灯光将色彩、物品和必要的功能置于前景之中，而不是简单地将灯光照在墙壁之上。其他规划项目则都依照阿西巴代姆医疗保健集团的要求设置。

1. The entrance of the building
2. A night view of the hospital
3. The façade has a lively and rich colour palette
4. The façade is made of structural glass

1. 建筑入口
2. 夜景中的医院
3. 建筑主体色彩活泼丰富
4. 建筑的表皮由结构玻璃墙组成

1

2

3

4

1

1. Delivery room
2. Neonatal ICU
3. Gallery
4. CVS ICU
5. Operating department

1. 产房
2. 新生儿加护病房
3. 画廊
4. 脑血管加护病房
5. 手术部

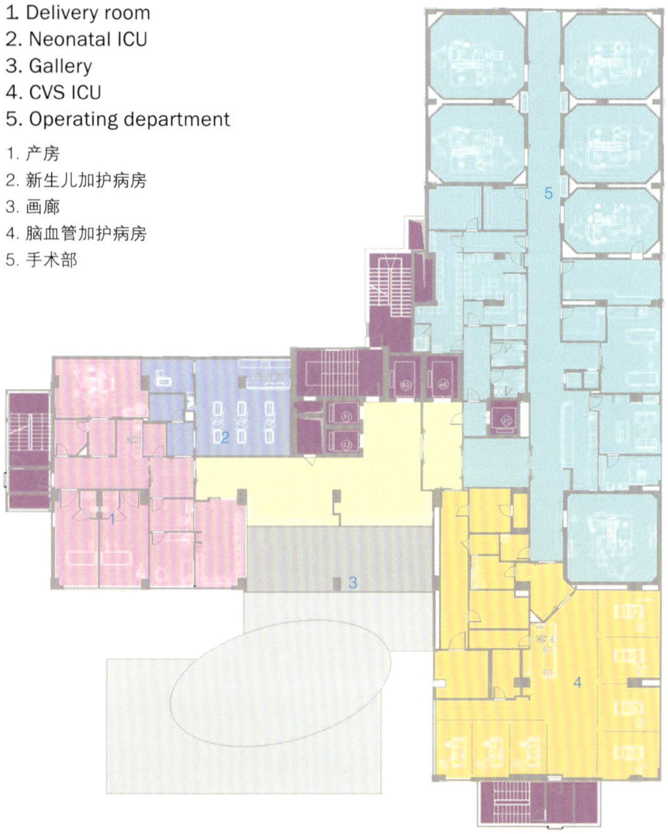

1. The unique ceiling of the central hall
2. The design of the hall is quite artistic

1. 中央大厅天花板独具特色
2. 大厅设计充满艺术感

2

1

1. Central waiting area
2. Reception
3. The layout of the patient room

1. 中央等候区域
2. 接待处
3. 病房布置

2

3

1

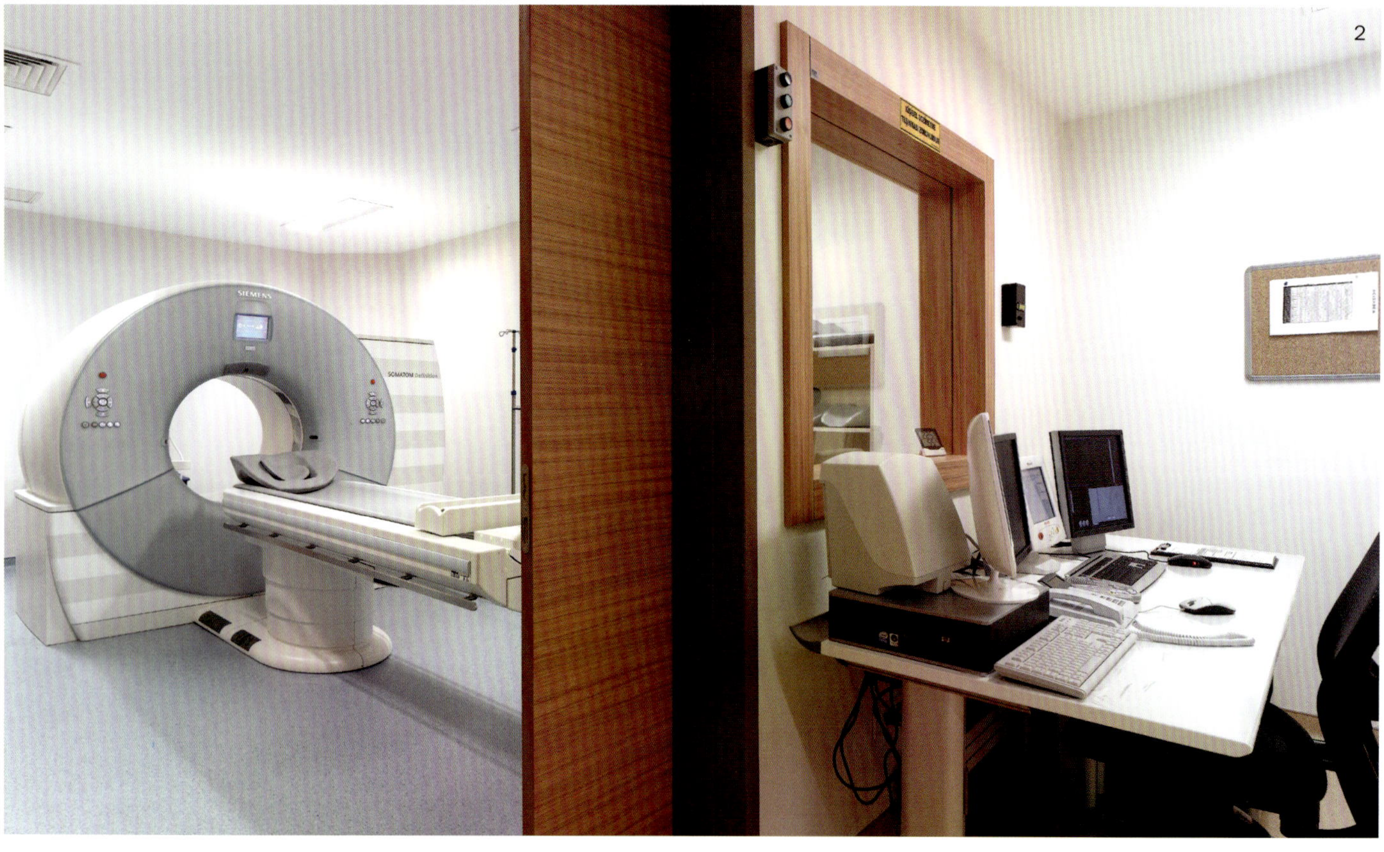
2

1. Common patient room
2. Medical area
3. The patient room is well-equipped

1. 公共病房
2. 医疗区
3. 功能齐全的病房

美因兹乔纳斯古腾堡大学保守医疗中心

Clinic of Conservative Medicine at the Johannes Gutenberg University in Mainz

Location: Mainz, Germany **Completion date:** 2007 **Designer:** Tönies+Schroeter+Jansen Frei Architekten GmbH **Photographer:** Johannes Voigt **Area:** 52,000 m^2

项目地点：德国，美因兹 完成时间：2007年 设计师：托尼斯+施罗特+詹森·弗雷建筑事务所 摄影师：乔纳斯·沃伊特 面积：52,000平方米

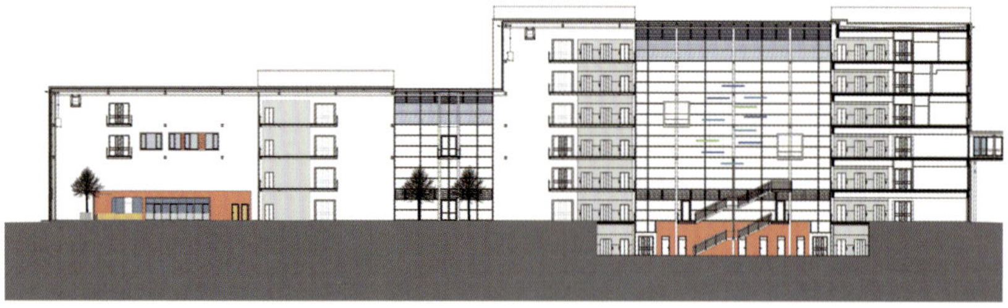

The new clinic for the department of conservative medicine at the Johannes Gutenberg University in Mainz is another step towards the modernisation of the entire hospital. A combination of different functions – such as internal medicine services, radiology and laboratories – required, made this building one of the largest projects in the history of the clinical centre.

This first construction stage realised on a plot of about 10,000 square metres within the clinics grounds resulted in a new hospital of 52,000-square-metre gross floor area with six floors and two basement levels, accommodating three main functional sections of the department. A main walkway – the "backbone" of the building – connecting the different health care services, provides a better workflow and therefore an outstanding environment for research and teaching.

The building is surrounded by an landscaped area with pergolas, water basins and semi-public courtyards open for visitors and patients alike, linking its inner space with the proximity. Rooms are facing green, enclosed and quiet areas, ensuring as much as fresh air and natural daylight as possible even for the basement. A value, optimising the use of space and helping patients to find their way around. The main walkway ("Magistrale") is fully glazed, opens to the green exterior and secures a supply of natural daylight for the whole building. This connection is the bonding part and makes a structured and logical conjunction possible.

The unique design and standard room dimension in height and depth, assures a flexible arrangement of functions in the future. This proves the sustainable approach of the building, allows for subsequent modifications and supported an efficient implementation design. Every building element interacts with the other and with the surrounding, by utilisation of different volumes and a complementary urban design. The use of modern materials, concerted proportions and a deliberate alignment of light and transparent building structures, creates an enduring impression.

A smart colour scheme – developed in cooperation with the artist Lee Kozlik, Lübeck – provides a pleasant atmosphere in every room, supports the recovering process of patients and maintains a lively working environment. The diverse requirements of patients and personnel demands different colour climates. For this the hospital is divided into small units with warm and cool zones. An arrangement which helps to orientate easily within the building as well; a familiarity on first sight.

Everything is connected by the main walkway: an interaction of light and shade, the colour of the sky and the interior materials, the time of day and different weather conditions. It creates a vivid building with subtle colour nuances.

保守医疗中心大楼是美因兹乔纳斯古腾堡大学临床中心迈向现代化的重要一步。内科、中心放射科和中心实验室等职能部门的结合让项目成为了临床中心历史上最大的建筑。

临床中心新建筑的一期工程阶段占地10,000平方米，总建筑面积为52,000平方米。保守医疗中心大楼分为3个主要部分，共有6层楼和2层地下室，分别容纳了不同的职能部门。一个主走道——“脊椎”连接了不同的医疗服务部门，提供了更好的工作流程，为研究和教学提供了优异的环境。

建筑四周由景观区域环绕，如藤架、水池和半公共庭院都吸引着患者和游客前来休息，增添了内部空间的亲近感。建筑与周围的环境紧密地联系在一起。房间朝向绿色、封闭而安静的区域，为室内空间（乃至地下室）提供充足的新鲜空气和自然光线。主走道采用全玻璃密封，通往绿色的室外，还保证了整座楼里的自然采光。这一连接线打造了一个有组织、有逻辑的连接结构。

1. The building is surrounded by natural landscape
2. The entrance
3. The design looks light and clear
4. A side view of the hospital

1. 建筑周围充满自然景观
2. 入口处
3. 建筑设计轻盈透明
4. 医院侧面

2

3

4

1

2

独特的设计和标准空间纵深比例让未来功能区的设置更加灵活。这保证了建筑的可持续性设计，便于后续的调整并支持了高效的设计实施。各个建筑元素在彼此之间和环境之间通过不同空间和现代城市设计形成了互动。现代材料、协调比例以及轻质而透明的建筑结构共同打造了令人印象深刻的持久形象。

项目的色彩运用由设计师与卢比克艺术家李·佐里克共同打造，巧妙的色彩搭配让每个房间都拥有愉悦的氛围，帮助患者尽快康复，也让工作人员健康舒适。患者与工作人员的不同需求衍生了不同的色彩氛围。医院建筑划分为冷暖色调的小单元，这让建筑的结构更加清晰，更具辨识度。色彩的识别提升了亲密感。

所有元素都与主走道相连：光影的互动、天空和室内材料的色彩、一天之中时间的变换和不同的天气条件。这是一座色彩斑斓的鲜活建筑。

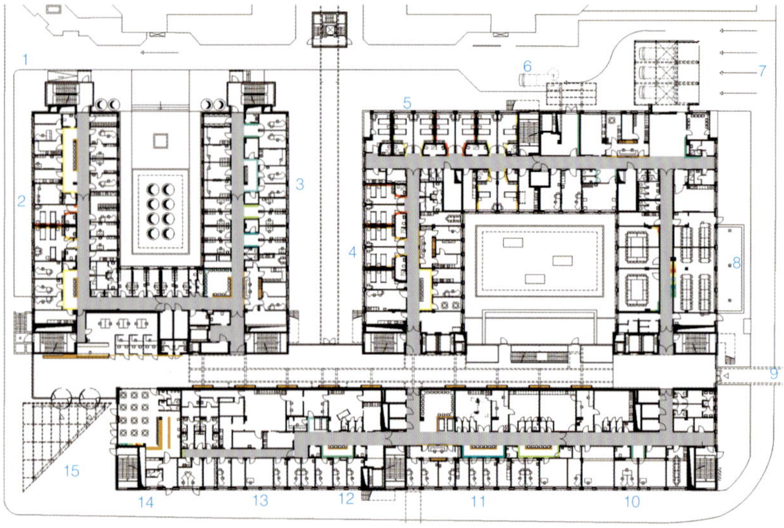

1. Entrance HIV policlinic
2. Medical day care unit HIV policlinic
3. Health care centre
4. Stat. 14 emergency ward
5. Stat. 13 emergency ward
6. Lay-by patients
7. Emergencies
8. Conference centre
9. Entrance employees
10. Radiology
11. Radiography
12. CT
13. Radiology
14. Café
15. Main entrance

1. 艾滋门诊入口
2. 艾滋门诊日间医疗护理病房
3. 保健中心
4. 14号急诊病房
5. 13号急诊病房
6. 病患停车处
7. 急救车停车处
8. 会议中心
9. 员工入口
10. 放射线科
11. X射线科
12. 断层扫描室
13. 放射线科
14. 咖啡厅
15. 主入口

1. The glass ceiling ensures good daylight
2. The design is bold and lively
3. Information centre

1. 玻璃顶棚保证了良好的自然采光
2. 设计大胆活泼
3. 信息台

1

2

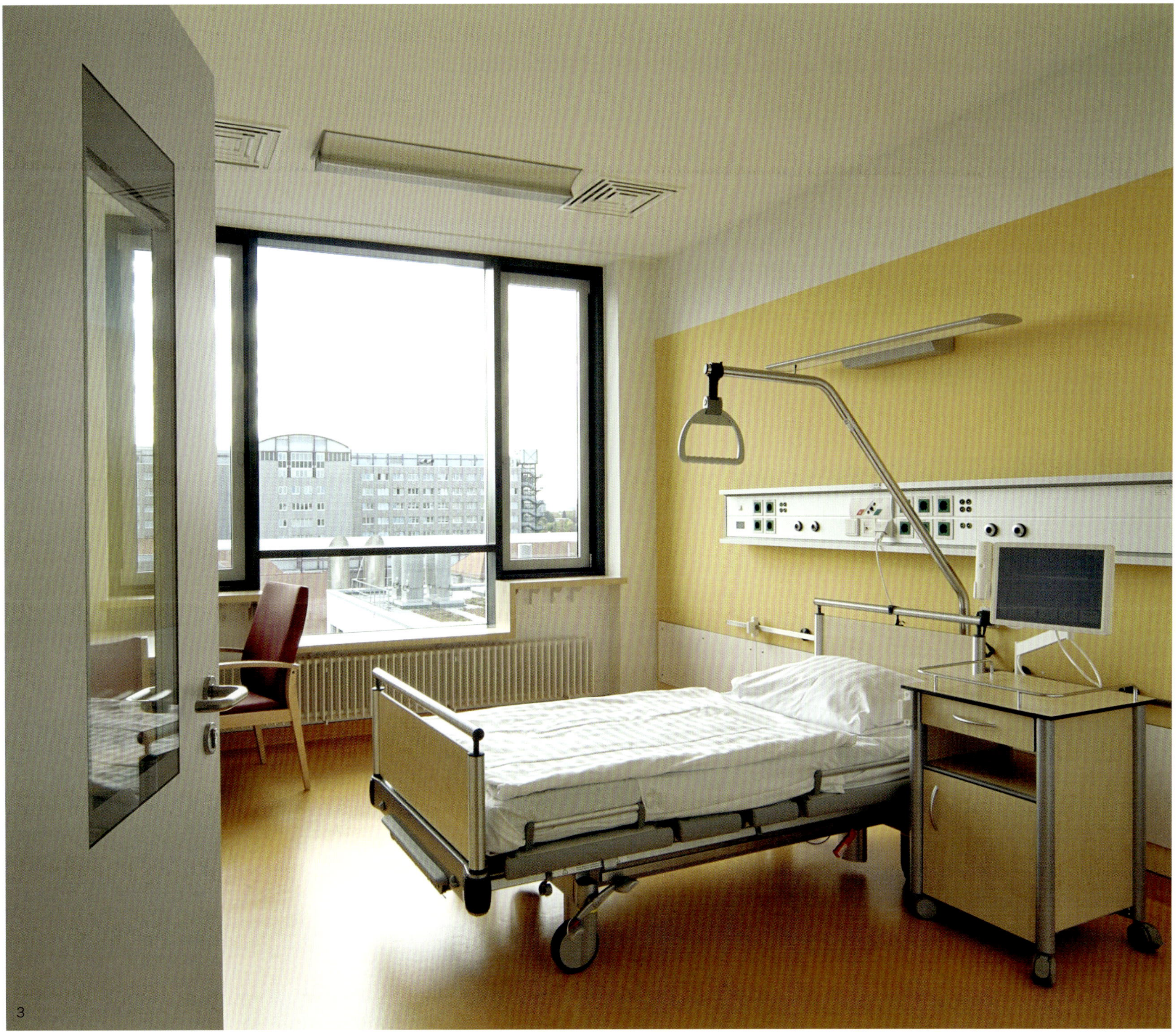

1. Reception
2. Medical area
3. The design of the patient room

1. 接待处
2. 医疗区
3. 病房设计

斯泰尔省级医院 Provincial Hospital Steyr

Location: Steyr, Austria **Completion date:** 2008 **Designer:** TEAM M Architekten ZT GesmbH **Photographer:** TEAM M Architekten ZT GesmbH **Area:** 44,010 m^2

项目地点：奥地利，斯泰尔 完成时间：2008年 设计师：M团队建筑事务所 摄影师：M团队建筑事务所 面积：44,010平方米

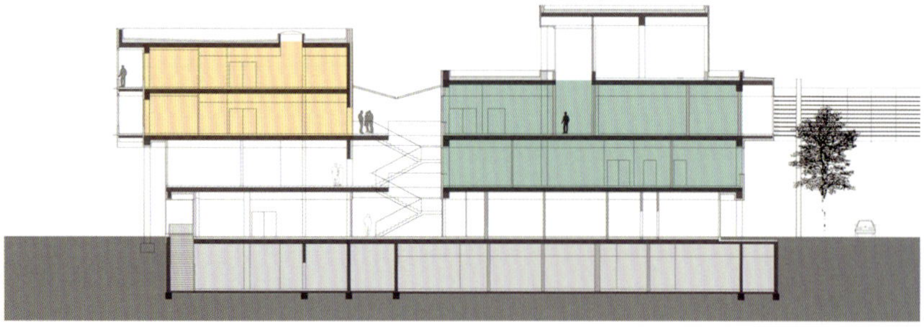

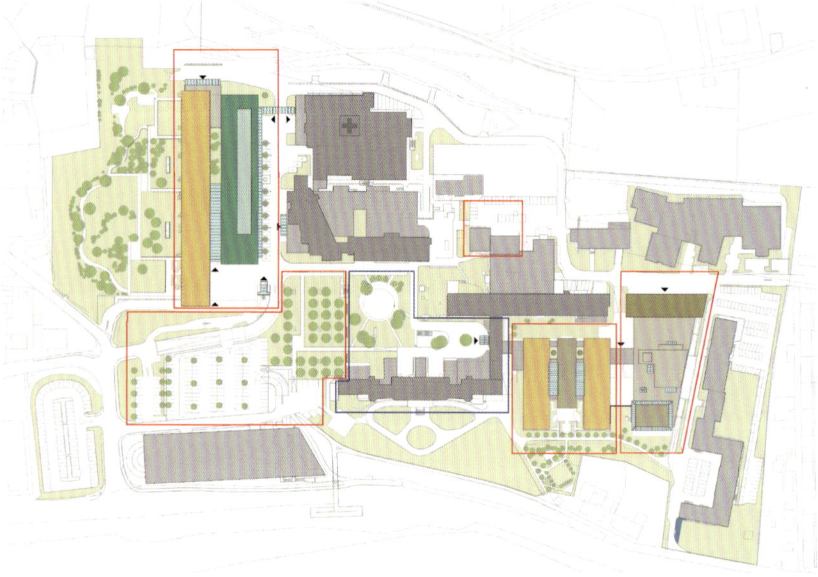

The Steyr Provincial Hospital is structured as a pavilion system. The various buildings were constructed between 1916 and the present as more space was needed, but there was no clear urban planning concept on which this expansion was based. New urban planning measures were taken to not only define a functional clear profile of the new buildings but also to integrate already existing construction axes and to respect existing building heights.

A clear strategy in architectural and urban planning terms was developed taking into consideration the heterogeneous structure of the existing buildings. The building structure now shows a clear north-south and east-west orientation. The "Surgery Centre" forms a clear termination of the building to the west and at the same time marks a transition to the generously laid out patients' garden; the administration building constitutes the termination towards the east. The urban planning strategy was continued, while giving clear expression to the architecture. Contrasts between full walls and transparent glass pieces, straightforwardness and clear structure are the principles governing the design.

As a matter of course several energy-technical considerations guided the design – considering prevention of overheating by means of structural measures (especially to avoid additional cooling in the patient rooms), reduction of outer surface of buildings and exploitation of passive solar energy (the glass façade of the administrative building and the glass parapets of the surgery centre provide 260-square-metre photovoltaic solar power plant).

In general the deliberate use of colour and material guided the design. For better orientation the designers have introduced clearly recognisable points such as the orange group of seats in front of the purple wall, etc. They tried to work with the colour in the material, which gives natural tones. Moving from one's own home to a hospital can be a critical event in a person's life, being linked as it is to a sense of being insecure, helpless and uprooted. A sensitive design of space and colour should contribute to the patients feeling secure, safe and at home.

The materials wood (in the bed wing) or brick (in the psychiatric ward) create a cosy atmosphere and a harmonious ambiance. They become an integral part of the project design since the hospital can only offer refuge and be a place for recovery when building material, ground plan, exterior space and surroundings merge to form one consistent structure.

斯泰尔省级医院由一系列的场馆组成。由于对空间的需求不断增加，从1916年至今，医院建造了许多建筑，但是这些扩建工程缺乏明确的规划概念。新的规划方式不仅为新建筑定义了明确的属性，还整合了原有的建筑结构轴线，并且尊重了它们的高度。

建筑和规划层面都将原有建筑形式各异的结构纳入了考虑之中。建筑呈现出朝向南北和东西的清晰结构。"手术中心"是建筑在西侧的终点，是建筑与花园之间的过渡。行政楼则是东侧的终点。延续的规划策略为建筑提供了清晰的表达方式。封闭式墙面和透明玻璃形成的对比、直截了当的建筑形式和清晰的结构是设计的主要原则。

设计采用了一些能源技术策略，通过结构措施避免了建筑过热（尤其是避免了病房的过度制冷），减少了建筑的外表面，并且采用了被动式太阳能（行政楼的玻璃外墙和手术中心的玻璃护墙提供了260平方米的光电太阳能动力装置）。

设计中，色彩和材料的运用都经过了深思熟虑。为了更好的定位，设计师采用了明显的色彩对比，如在紫色的墙壁前设置一排橘黄色的座椅等。他们试图让色彩与材料结合在一起，形成自然之感。从家里搬到医院有可能是人生的重大事件，会让人感到不安、无助和没有归属感。敏感的空间和色彩设计将让患者感到安全、可靠和家一般的舒适。

木材（病房楼）和砖块（精神科病房）营造出舒适、和谐的氛围。他们成为了项目设计必不可少的部分。当建筑材料、平面规划、室外空间和周边环境结合在一起，医院将成为患者的避风港，帮助他们恢复健康。

1. The façade of the hospital
2. The main volume is made of red bricks
3. The large glass windows provide abundant daylight

1. 医院正面投影
2. 棕红色砖块构成的建筑主体
3. 大型玻璃窗镶嵌在建筑墙体上保证了采光

1

2

1

2

1. The architecture is connected with the garden
2. The architecture takes white as its main colour
3. A side view of the building

1. 建筑与花园相连接
2. 白色为主色调的建筑主体
3. 建筑侧立面

3

1. Reception
2. Consulting room
3. Conference room
4. Café
5. Hall
6. Dining room
7. Kitchen

1. 接待处
2. 诊疗室
3. 会议室
4. 咖啡厅
5. 大厅
6. 餐厅
7. 厨房

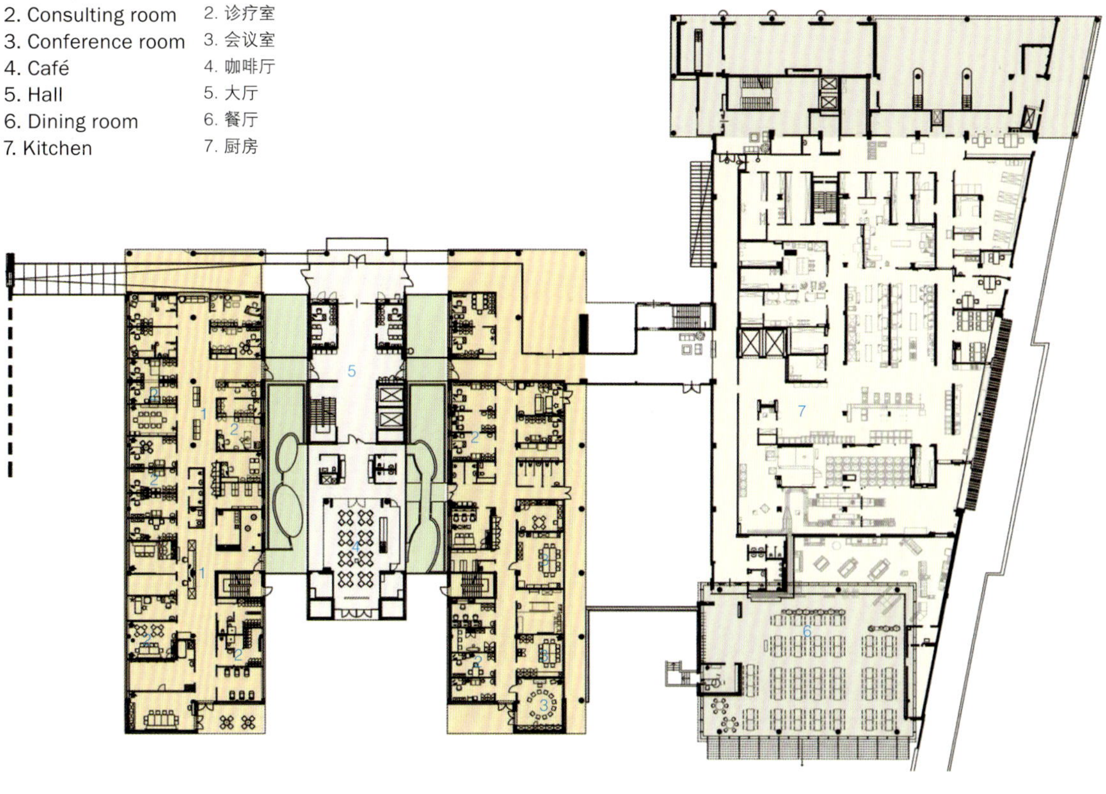

1

1. The interior landscape
2. Details of the landscape

1. 建筑内部的自然景观
2. 景观细节

2

1

2

3

1. Dining area
2. The spacious and open space
3. Public area

1. 就餐区
2. 宽敞开阔的空间
3. 公共区域

皇后医院中心大楼 The Pavilion at Queens Hospital Centre

Location: New York City, USA **Completion date:** 2006 **Designer:** Perkins Eastman Architects **Photographer:** Paul Rivera/ArchPhoto **Area:** 13,378 m^2

项目地点：美国，纽约 完成时间：2006年 设计师：珀金斯・伊斯曼建筑事务所 摄影师：保罗・里韦拉建筑摄影 面积：13,378平方米

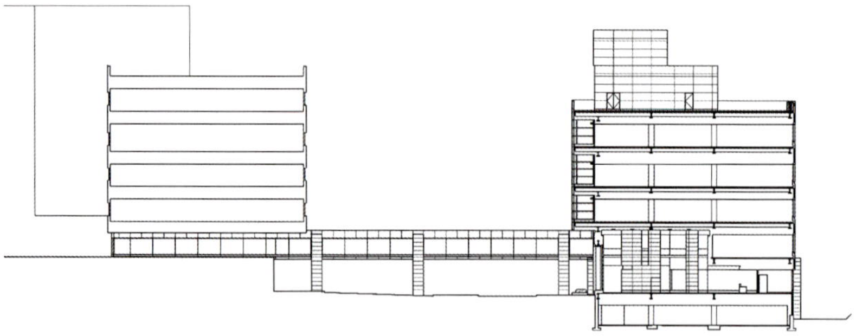

The Pavilion at Queens Hospital Centre represents the culmination of an extensive project to replace one of New York City Health and Hospitals Corporation's oldest facilities. Taking into consideration the specific needs of this vibrant New York City borough, the Health and Hospital Corporation devised a master plan that brings to Queens a medical facility equipped to support its community well into the 21st century – as well as creating a signature landmark for the hospital campus at its northern edge.

With the divestment of properties and demolition of outmoded facilities, this project reassembles essential healthcare and administrative components in a more compact, convenient, and cost effective adjunct to the Queens Hospital Centre. The Pavilion houses primary care, pediatrics, behavioral health, dental, diabetes, and eye care functions; in addition to housing administrative, educational, and training facilities for the entire hospital centre.

In order to accommodate all of these expectations, the planning process adhered to design principles that included maintaining separation of patient areas from staff areas, providing separate access for patients and staff, maximising the use of natural sunlight wherever possible, placing patient circulation along day-lit corridors, creating memorable spaces to facilitate wayfinding, planning for operational efficiencies, and building in flexibility for the future.

The elegant five-level building is organised by a 300-foot-long glazed public concourse along the southern perimeter. This cantilevered circulation spine links all clinics to the staff and service elevator core on the west and the public elevator core to the east. Composed of pre-cast concrete and a glass curtain wall, the building also provides a light-filled two-storey atrium and public entry plaza at the eastern end and a staff/service entry on the west.

In addition to its functional requirements, the Pavilion exceeds the goals of welcoming outpatients in a dignified and life-affirming context; being operationally efficient; accommodating changing functions over time; attracting physicians and other practitioners; reducing patient anxiety levels and staff stress levels; serving as an attractive and inviting entry point for the entire hospital centre.

皇后医院中心大楼代表了医疗建筑的最高峰，取代了纽约健康医疗公司的旧医疗设施。健康医疗公司在考虑到纽约不断增长的医疗需求后，计划为皇后区打造一座21世纪医疗建筑，同时也为医院在北端打造一座地标性建筑。

通过资产的转让和旧设施的拆除，项目将基本医疗服务和行政元素重新组合在更紧凑、便利和高效的附属建筑之中。大楼里设有基层医疗、儿科、行为健康科、牙科、糖尿病科和眼科设施，还包含了整个医疗中心的行政、教育和培训空间。

为了将这些设施结合在一起，项目所采用的基本原则包括：分离患者区和工作区、为二者提供独立的通道、最大化自然采光、将患者通道设在自然照明走廊内、打造明显的指路区域、提高运作效率和为未来改造提供灵活性。

优雅的五层建筑围绕着南侧一座91米高的玻璃幕墙公共广场展开。悬臂式走道将各个科室和西侧的员工/服务电梯和东侧的公共电梯连接起来。建筑由预制混凝土和玻璃幕墙组成，还在东侧设有一个两层楼高的中庭和公共入口广场，在西侧设有一个员工/服务入口。

除了功能需求之外，大楼还为门诊病人提供了严肃而令人安心的环境，能够有效地进行运营，随着时间的变化而变换功能，吸引医生和其他实习者，减少患者的焦虑水平和工作人员的压力，为整个医疗中心提供具有吸引力的入口。

1. A side view of the building
2. Details of the façade
3. The façade is made of concrete and glazed wall

1. 建筑侧面投影
2. 外立面细节
3. 混凝土与玻璃幕墙共同构成了建筑表皮

1

2

3

1. A graceful five-storey building
2. The curtain wall is transparent and bright

1. 优雅的五层建筑
2. 玻璃幕墙通透明亮

1. Waiting
2. Md workstation
3. Chart storage workroom
4. Director
5. Therapy/patient Ed.
6. Psychologist
7. Visit prep.
8. Case Worker
9. Conference
10. Phlebotomy
11. Testing

1. 候诊区
2. 医疗站
3. 图表仓库工作室
4. 主任室
5. 治疗/患者病房
6. 心理学科
7. 治疗准备室
8. 病例研究
9. 会议室
10. 刺骼疗法室
11. 测试室

2

1 The open central hall
2. Interior space is calm and peaceful

1. 开阔的中央大厅
2. 内部空间氛围平静祥和

2

1

2

1. Medical area
2. Corridor
3. The resting area with interesting wall paintings
4. The comfortable and clean waiting area

1. 医疗功能区
2. 走廊设计
3. 休息区童趣十足
4. 舒适整洁的等候区

3

4

阿斯塔医疗中心 Assuta Medical Centre

Location: Tel Aviv, Israel **Completion date:** 2009 **Designer:** Zeidler Partnership Architects **Photographer:** Tom Arban **Area:** 60,000 m²

项目地点：以色列，特拉维夫 完成时间：2009年 设计师：扎德勒建筑事务所 摄影师：汤姆・阿本 面积：60,000平方米

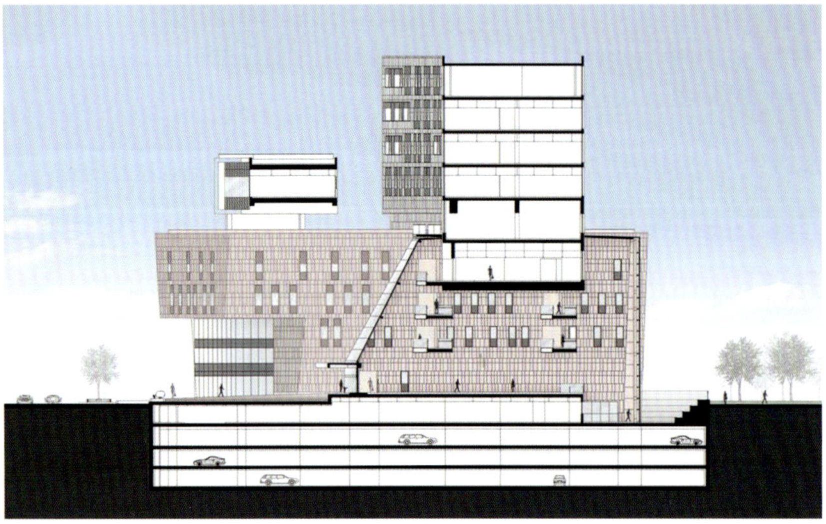

The firm was chosen by the Assuta Medical Centrexun to design a 400-bed, 600,000-square-foot diagnostic and surgical procedures hospital in Tel Aviv. One of the key challenges in designing the new hospital was connecting the Park on the south side and Habarzel Street to the north. This connection is achieved by cutting the north-south axis of the hospital, perpendicular to the park, in two places with a large glass concourse. The resulting public space creates a link between the interior healing spaces of the hospital and exterior site but also forms the central spine, guiding visitors and patients through the hospital. This path is reinforced by an angled cut at the bottom of the building mass, directing circulation from the east plaza to the main entrance of the building. The cantilevered mass not only encourages people to move towards the front entrance but also provides shelter.

The hospital site presented a second major design challenge: initially developed as part of a pre-existing two-phase commercial development, the site was excavated with all the foundations in place to receive an office complex. Designing the procedural/diagnostic hospital programme on the existing foundations and taking the client's objectives into account resulted in a unique design appropriate to the facility. In order to achieve this, a variety of tools were used to enable the client group to make informed decisions at each step of the design process.

Designing an international healthcare facility requires an understanding of the site and the local culture in order to understand the nature of the project. The design process for the Assuta project successfully merged patient expectations with the hospital's objectives by integrating the site, culture and medical systems. The hospital's goal is to provide a new standard of treatment and services currently unknown in Israel. The facility will attract clients and the best medical staff allowing for sustainable future growth.

The client group consists of an insurance company that will manage the hospital, providing both the medical and non-medical services and the development company, which constructed and owns the building. This unique joint venture had a profound influence on the design of the building. The design of this facility reflects the integration of technology, cultural spaces, and the environment down to even the smallest details helping to create a unique medical experience.

The Assuta Medical Centre's main services are invasive and non-invasive, diagnostic and procedural. They include, in phase one of a two-stage construction process, outpatient and inpatient facilities with 16 Operating Rooms, 27 Intensive Care units, 230 acute care beds, 40 day-surgery beds, Imaging, Catheter Labs, Radiation Therapy and 25 Dialysis Positions. The new centre will add significantly to the capacity of the existing 6 Assuta facilities in Israel.

阿斯塔医疗中心选择扎德勒建筑事务所在特拉维夫设计一个拥有400个床位、总面积55,742平方米的诊断和外科手术医院。医院设计的主要挑战之一是连接南面的公园和北面的哈巴尔泽尔街。这一连接通过医院垂直于的南北轴线得以实现。中间的公共空间在医院的室内治疗空间和外部场地之间建立了联系，也形成了一个中央走道，引导着访客和病人穿过医院。这个走道通过建筑底座的变相切入得到了加强，将东广场的人流引到主入口。这个悬臂式空间不仅引导着人们走入医院，还在广场和建筑之间提供了遮蔽物。

医院场地是第二个主要设计挑战：建筑场地最初是作为两个阶段的商业项目开发，地基设计全部以办公楼为基础。在原有的地基上设计外科/诊断医院项目并且实现客户的目标形成了一座独特的设计。为了达到这一需求，设计师运用了多种工具，让客户们在每一步设计中都享有知情权。

1. Habarzel Street entrance view
2. The glazed wall is transparent and bright
3. View from Habarzel Street

1. 哈巴尔泽尔街入口风光
2. 玻璃墙体通透明亮
3. 从哈巴尔泽尔街看建筑

3

1

设计一座国际化医疗设施要求设计师对场地和当地文化进行全面的理解，彻底了解项目的本质。阿斯塔医疗中心的设计成功地通过场地、文化和医疗系统的结合，将病患的期望值和医院的目标融合在一起。医院的目标是提供以色列现在所没有的高标准治疗和服务。医疗中心在未来将吸引更多的客户最好的医疗工作者。

项目的客户群由管理医院、提供医疗与非医疗服务的保险公司和建造并拥有大楼的开发公司组成。这一独特的联合经营对建筑的设计意义深远。项目设计反映了技术、文化空间、环境和细节的结合，打造了独特的医疗体验。

阿斯塔医疗中心主要服务既有诊断又有手术。一期工程包括门诊和住院设施，共有16间手术室、27间特护病房、230个急症护理床位、40个日间手术床位、成像室、导管实验室、放射治疗室和25个透析床位。新中心将为阿斯塔现有的6个医疗设施增添更多功能。

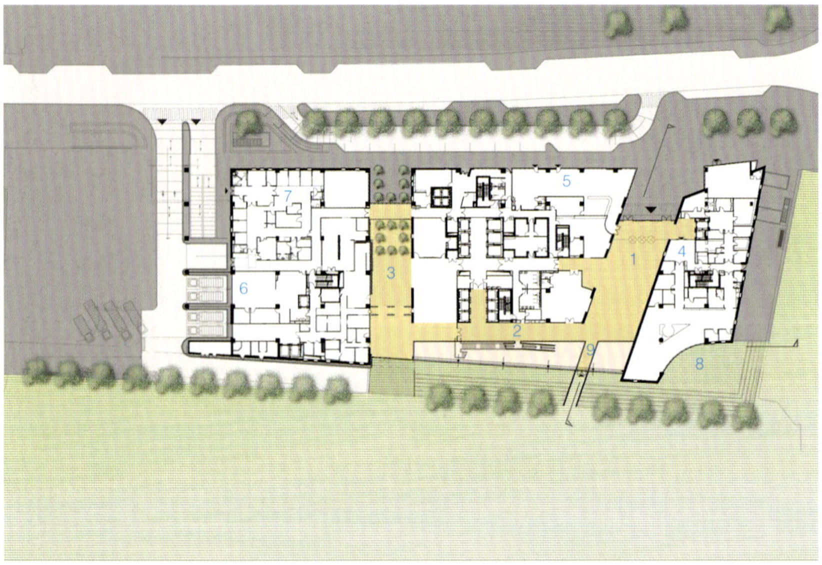

1. Main entrance
2. Atrium
3. Cafeteria
4. Dialysis
5. Retail and education
6. Receiving and services
7. Walk-in clinic
8. Sunken garden
9. Link to the park

1. 主入口
2. 中庭
3. 自助餐厅
4. 透析室
5. 零售和教育空间
6. 接待和服务区
7. 门诊
8. 下沉花园
9. 公园通道

2

1. The building mass is composed of two bodies – a lighter white mass resting on a heavy, reddish brown mass
2. The entrance to the hospital

1. 建筑由两部分组成——轻质的白色结构坐落在厚重的红棕色结构上
2. 医院入口处

1

1. The interior colour is rich and abundant
2. The high glass wall provides enough daylight for the interior

1. 医院内部色彩丰富
2. 高大的玻璃墙保证了室内的自然采光

2

1

2

3

1. Chapel view from the atrium
2. View of atrium stairs
3. View of atrium from upper stairs

1. 从中庭看小礼拜堂
2. 中庭楼梯
3. 从上层楼梯看中庭

肯尼迪·克里格研究所门诊中心 Kennedy Krieger Institute Outpatient Centre

Location: Baltimore, USA **Completion date:** 2009 **Designer:** Stanley Beaman & Sears **Photographer:** Jim Roof Creative **Area:** 10,683 m²

项目地点：美国，巴尔的摩 完成时间：2009年 设计师：斯坦利·毕曼&希尔斯 摄影师：吉姆·卢夫创意公司 面积：10,683平方米

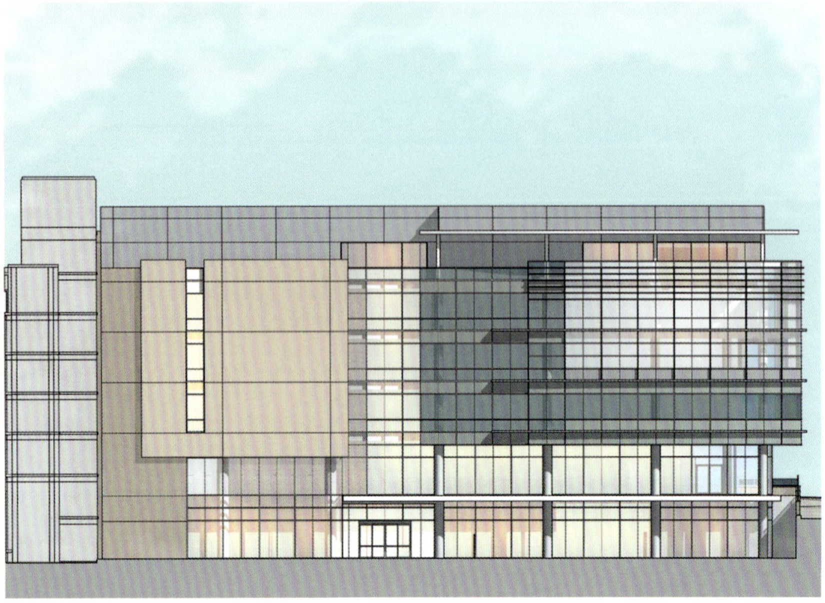

Outpatient services of the Kennedy Krieger Institute, which is dedicated to research, education and therapeutic care for children with neurological injuries and developmental disabilities, were housed in scattered locations. The Harry and Jeannette Weinberg Building consolidates these functions into a structure in which building, interior and landscape design merge to promote dignity, interaction, independence and the exploration of potential.

The design team conceived the building symbolically as a three-part process. Therapy, or body, is expressed in its precast and mass. Research, or mind, is evoked by its curtain wall and abundant natural light. And hope, or spirit, is conveyed by the transparent interface of the building's interior with its therapy garden and, from upper floors, panoramic city views.

The principal public space is a two-storey lobby, oriented towards the north and providing gentle natural light. Outside it, the block-long garden is organised into distinct "rooms" devoted to mobility activities for young patients and respite for families and staff. Located along the building's front, but recessed and protected by a low wall, its design respects patients' and families' privacy needs without hiding the Institute's presence. An arcing path that commences at a water feature leads into the lobby with the gestural curve of a "welcome wall". This illuminated billboard articulates the theme of body, mind and spirit through silhouetted imagery of active children and the inspirational slogan "In my mind, I can do anything."

Arriving patients and families are greeted by the lobby's organic shapes, nature-themed colour palette and light-reflecting, touch-inviting materials. These strategies continue throughout the building, helping to overcome dispiriting stereotypes of therapeutic facilities, as do the structure's pervasive transparency and natural light. Meanwhile, its modernist aesthetic communicates the Institute's commitment to research, state-of-the-art medicine and the possibilities of healing.

肯尼迪·克里格研究所门诊服务包括儿童神经创伤和发育性残障的研究、教育和治疗。哈利和珍妮特·温伯格楼集中了这些功能，将建筑、室内和景观设计结合在一起，以尊重、互动、独立和潜力探索为主要理念。

设计团队将建筑看成三个部分。治疗部分（身体）以混凝土模块来表示；研究部分（精神）以幕墙和充足的自然光为特色；希望（心灵）部分则由建筑内部的透明界面来传达，可以看到治疗花园和城市全景。

主要公共空间是一个朝北的双层大厅，提供了充足的自然光。大厅外的花园被划分为独立的空间，为年轻的病患提供活动空间，也为家属和工作人员提供休息场所。建筑正面有一面矮墙，它的设计保证了病人和家属的隐私，又不会遮挡建筑的外观。由水景边延伸的弧形走道通往大厅，沿线是一面弧形"欢迎墙"。这面灯光布告板通过象征性的标语"在心里，我可以做任何事"连接了身体、精神和心灵。

到达研究所的病患和家属将被大厅有机的造型、自然系的色彩和反光、有触感的材料所吸引。这些策略在建筑内部延伸，颠覆了治疗机构的固定模式，为建筑带来了风景和自然光。同时，它现代、美观的造型传达了研究所在研究、医疗以及康复潜力方面所做的努力。

Award:

2010 ASID Design Excellence Award – Healthcare

2010 IIDA Best of the Best Awards – Healthcare

获奖情况：

2010年ASID卫生保健方向杰出设计奖

2010IIDA卫生保健方向最优奖

1. The name of the therapy garden is shown
2. Garden access in front of the building
3. The main façade is made of glazed curtain wall

1. 疗养花园名称展示
2. 楼前的花园小径
3. 建筑主体采用玻璃幕墙

3

1

2

1. The modern and innovative architecture form
2. The garden in front of the hospital could provide space for patients and family to rest and play
3. A side view of the building

1. 建筑造型现代新颖
2. 医院前方建有花园景观供患者和家属休闲游览
3. 建筑侧面

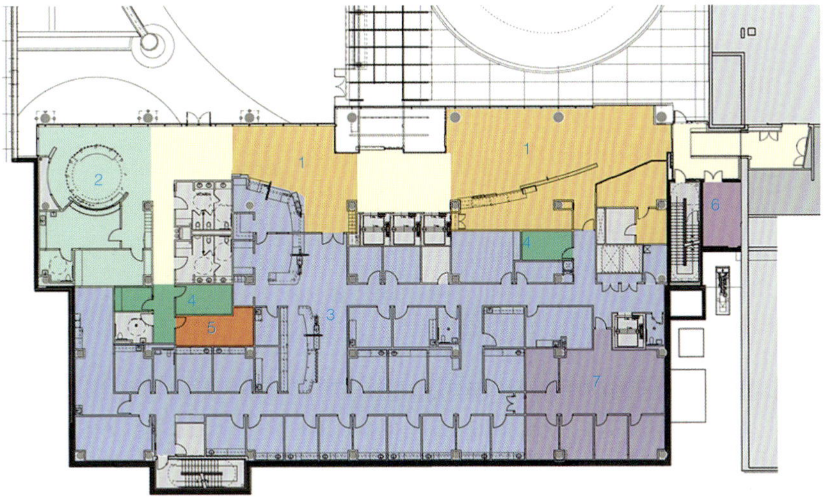

1. Reception & public area
2. Family/child waiting
3. Outpatient clinic
4. Guest relations
5. Resource centre
6. Facility
7. Outpatient offices

1. 接待处和公共区
2. 家属/儿童等候区
3. 门诊治疗区
4. 客人关系室
5. 资料中心
6. 辅助设施
7. 门诊办公室

1

2

3

4

1. The hall design is quite artistic
2. Reception
3. The bright and open space
4. The interior layout is simple and bright

1. 大厅设计艺术感十足
2. 接待处
3. 空间明亮开阔
4. 室内布置简单明快

1. The purple office area is elegant and peaceful
2. Reception
3. The interior is elegant and generous
4. Children's therapy gym

1. 紫色的办公区高雅宁静
2. 接待台
3. 室内造型美观大方
4. 儿童治疗健身室

3

4

雷乌斯圣约翰医院 Sant Joan de Reus Hospital

Location: Reus, Spain **Completion date:** 2011 **Designer:** Corea Moran Arquitectura S.L., Pich-Aguilera Architects **Photographer:** Corea Moran Arquitectura S.L. **Area:** 92,073 m²

项目地点：西班牙，雷乌斯 完成时间：2011年 设计师：科里亚・莫兰建筑事务所、皮克–阿奎莱拉建筑事务所 摄影师：科里亚・莫兰建筑事务所 面积：92,073平方米

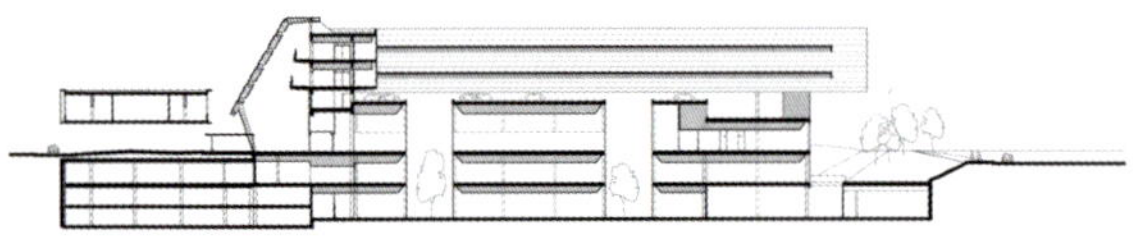

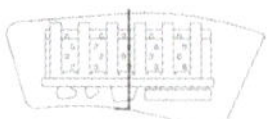

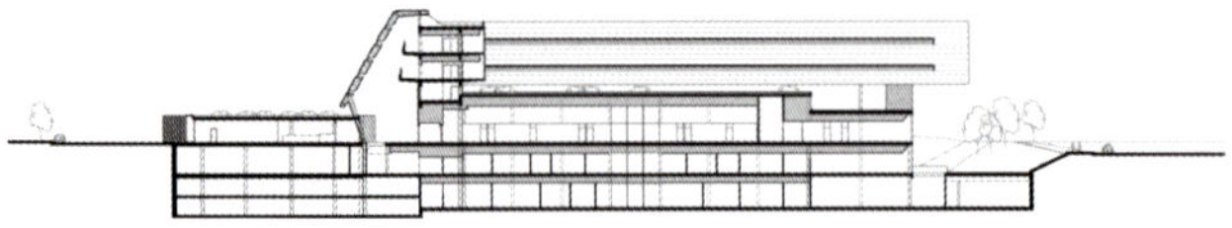

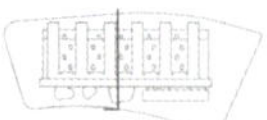

This project is presented according to the urban requirements of the plot and the guidelines established by the functional programme. It must be taken into account that the building, due to its public use and size, is a very important benchmark for the new technological park and the city of Reus. At the same time, the facilities are going to give an emblematic and modern image so as to achieve a high architectural quality and efficiency as a new hospital.

The objective of the new hospital building project is to take the most of the site features by improving the functional programme and its link between the different areas and their surroundings, and also optimising the vehicular traffic and flow of people. After evaluating different alternatives, the project is presented as a major horizontal unit with light wells on which six two-storey hospitalisation volumes rest as if they were floating. These volumes are linked to a large public circulation area; the slanting facade and the project of some smaller annex buildings give the hospital a more pleasant appearance along with the city, the technological park and the new university campus.

The project consists in building a low height hospital, with changes in the existing topography, a big well-lit unit where the general public and the physicians can communicate, with green spaces, evoking nature in the building area. In the facade, where pedestrian accesses are located, the large scale building is softened with two strategies.

On the one hand, the facade plane that covers the whole building is bent over, optically reducing the physical presence of the building and increasing its perspective effect. On the other hand, on the ground floor there are rounded pavilions at a lower scale that receive people, helping to organise the different entrances. In the big public area, the personnel, patients and visitors encounter a public square which enters into the building as if it were part of it.

项目展示了场地的城市需求和功能项目的指导方针。鉴于建筑的公共用途和规模，它将成为新科技园和雷乌斯城重要的坐标。同时，医院设施还将成为具有象征意义的现代形象，达到高标准的建筑品质和新医院的要求。

医院项目的设计将最大化地凸显场地特色，提升项目功能，促进各个区域与周边的联系，同时也优化垂直交通和人流。在评估了各个备选方案之后，项目以带有光井的水平单元为底座，上方是类似悬浮在空中的六层双层楼高医院空间。这些空间与大型公共流通区域相连，倾斜的外墙和较小的附属构筑物为建筑提供了更加愉悦的外观，在城市、科技园乃至新校园都独树一帜。

项目由一个随着地势而变换的低层医院、一座巨大的天光单元（医务人员和民众可以在此互相交流）和唤醒自然的绿化空间组成。入口设在建筑正面，项目柔化了建筑的巨大造型。

一方面，覆盖整座建筑的平板向内弯曲，在视觉上减少了建筑的实体存在感，增强了它的远景效果。另一方面，一楼的圆形接待大厅还能将各个不同的入口组织起来。在大型公共区域，工作人员、患者和访客将见到一个融入建筑内部的公共广场。

1. Entrance of the hospital
2. White exterior wall
3. The hospital has a unique appearance
4. Side view of the building

1. 医院入口
2. 白色建筑外墙
3. 医院外部造型独特
4. 建筑侧面

3

4

1

1. A hall full of light
2. Corridor on the first floor
3. The sophisticated interior decorations

1. 大厅采光良好
2. 二楼走廊
3. 建筑内部设计精致

1. Entrance	1. 入口
2. Hall	2. 大厅
3. Cafeteria	3. 餐厅
4. Chapel	4. 礼拜堂
5. Convention hall	5. 会议厅
6. Admission	6. 入院许可处
7. Administration	7. 行政部
8. Outpacients	8. 门诊部
9. Rehabilitation	9. 康复区
10. Day hospital	10. 日间医院
11. Detox area	11. 戒瘾区
12. Archive	12. 档案室
13. Medical offices	13. 医药室

1

2

1. Rehabilitation room
2. Office
3. Patient room
4. Waiting area

1. 康复室
2. 办公室
3. 病房
4. 等候区

圣玛丽医院和区域医疗中心

St. Mary's Hospital & Regional Medical Centre

Location: Grand Junction, USA **Completion date:** 2010 **Designer:** Perkins + Will **Photographer:** Ken Redding **Area:** 7,023 m² renovation + 40,784 m² new construction

项目地点：美国，大章克申 完成时间：2010年 设计师：珀金斯+威尔 摄影师：肯·雷丁 面积：7,023平方米翻修+40,784平方米新建

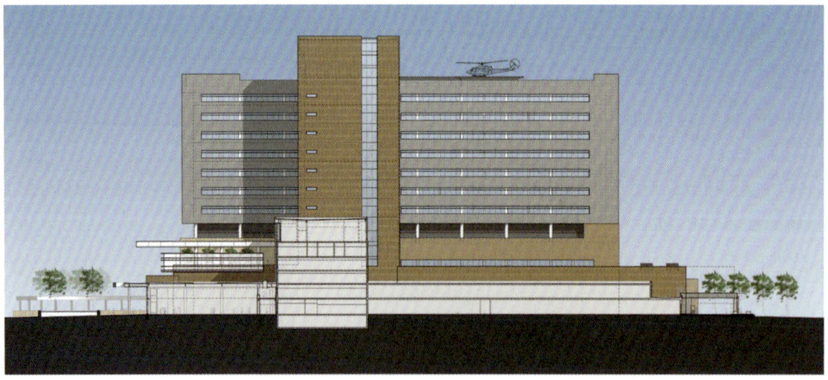

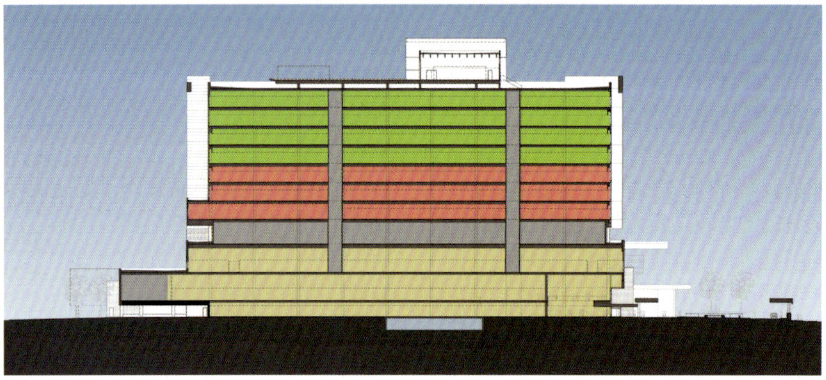

The current Century Project at St. Mary's Hospital and Regional Medical Centre, located near the Grand Junction airport, is intended to provide the latest medical care and technology to the residents of the Western Slope of Colorado and eastern Utah. A multi-year master plan will guide phased renovation + expansion and accommodate new technologies to meet the future healthcare needs of the growing community.

The multi-phase project is anchored by the new 12-storey patient bed tower. The project involved a new and renovated emergency department, operating rooms and a variety of inpatient beds, all accomplished while St. Mary's provided full hospital operations. The new tower opened in early 2010 and has achieved LEED Silver certification.

The architects wanted the new tower to complement the existing hospital – as well as Grand Junction's rocky landscape – while establishing a more advanced, forward-looking image. To avoid competing with the surrounding mesas and cliffs, the Perkins+Will team chose a simple rectangular silhouette for the tower. They selected brick and glass for the lower floors of the project – the base of the tower and an adjoining one-storey wing housing the cafeteria – to echo the scale and the brick exterior of the existing hospital. Once the tower clears the height of the original building, the cladding transitions to aluminium panel and glass for a more weightless, contemporary look. With patient rooms set on the long sides of the rectangle, the tower is oriented to maximise views to nearby landmarks (the Grand Mesa and the national forest to the south, the Book Cliffs to the north), while the façade's green glass recalls the local sagebrush. With sustainable features that include aluminium cladding made largely from recycled content, a weather-controlled irrigation system, and metres that monitor energy consumption.

圣玛丽医院和区域医疗中心临近大章克申机场，为科罗拉多西坡和犹他州东部的居民提供了最新的医疗服务和技术。计划多年的总体规划将为阶段性翻修和扩建进行指导，运用最新科技来满足不断发展的社区未来的健康需求。

分阶段项目以新建的12层病房楼为标志，包括全新的急诊部、手术室和各式住院病房。新病房大楼于2010年初开放，获得了美国绿色建筑协会银奖认证。

建筑师试图让新建的病房楼完善原有的医院结构和大章克申的岩石景观，同时建立一个更先进、更具前瞻性的形象。为了避免与周边的平顶山和悬崖形成冲突，珀金斯+威尔设计团队选择了简单的长方形造型。他们采用砖石和玻璃打造建筑的底层——塔楼的底座和连接的单层食堂与原有医院的规模和砖墙遥相呼应。在塔楼超过原有建筑之后，上层的外墙便以铝板和玻璃为主，营造出轻盈、现代的外观。病房设置在长方形的两个长边上，建筑的设计将酒店视野最大化——南邻大平顶山和国家公园，北临布克悬崖。外墙的绿色玻璃让人想起当地的艾蒿植物。建筑的可持续特征包括回收铝板、天气控制灌溉系统和能源消耗测量系统。

1. A long-range perspective of the hospital
2. A general view of the building
3. The upper façade is mainly made of aluminium sheet and glass

1. 医院远景
2. 建筑纵览
3. 建筑上层外墙以铝板和玻璃为主

Award:

2010 Healthcare Award, IIDA Northland Chapter FAB Awards

获奖情况：

2010医疗保健奖；国际室内设计协会北部地区分会FAB大奖

1

2

3

1

2

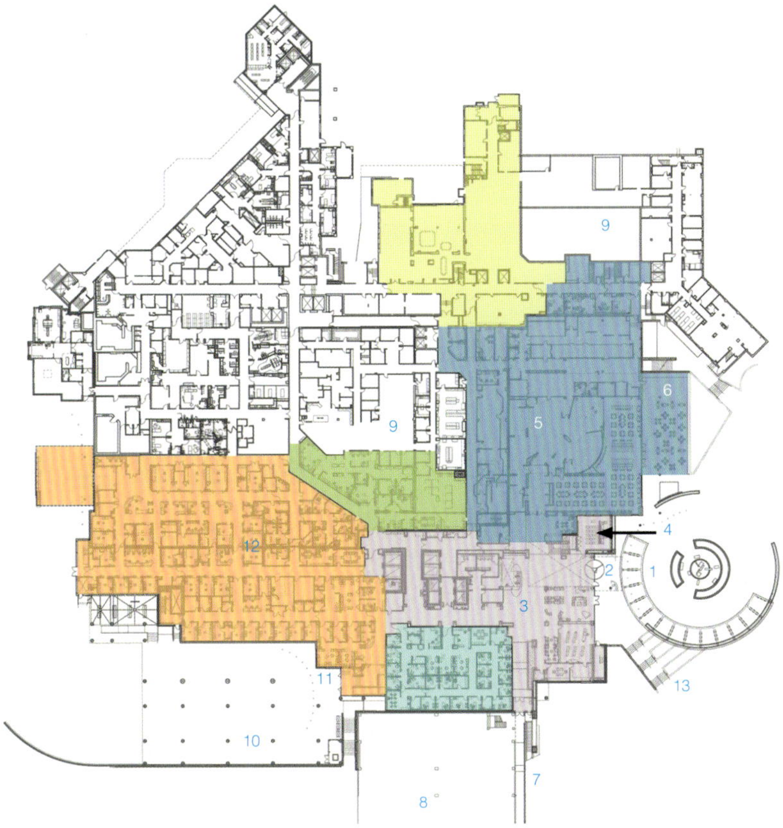

1. Spacious hall
2. Waiting area

1. 大厅空间宽敞
2. 等候区

1. Main entry/drop off
2. Main hospital entrance
3. Lobby
4. Reflection room
5. Cafeteria/servery
6. Cafeteria patio
7. Pedestrian link
8. Parking garage
9. Existing hospital
10. Emergency parking
11. Emergency entrance
12. Emergency/trauma
13. Monumental entry stair

1. 大门/下车口
2. 医院主入口
3. 前厅
4. 映像室
5. 餐厅/备餐室
6. 餐厅露台
7. 人行道
8. 停车场
9. 既有医院建筑
10. 紧急停车场
11. 紧急入口
12. 急诊室
13. 入口楼梯

2

3

1. The elegant corridor design
2. Restaurant
3. Medical treatment area
4. Reception

1. 走廊设计典雅
2. 餐厅
3. 医疗区
4. 接待处

4

马尔海济马丽莎夫养老院 Marishof in Maarheeze

Location: Maarheeze, the Netherlands **Completion date:** 2007 **Designer:** UArchitects **Photographer:** Norbert van Onna

项目地点：荷兰，马尔海济 完成时间：2007年 设计师：U建筑事务所 摄影师：诺波特・凡・翁那

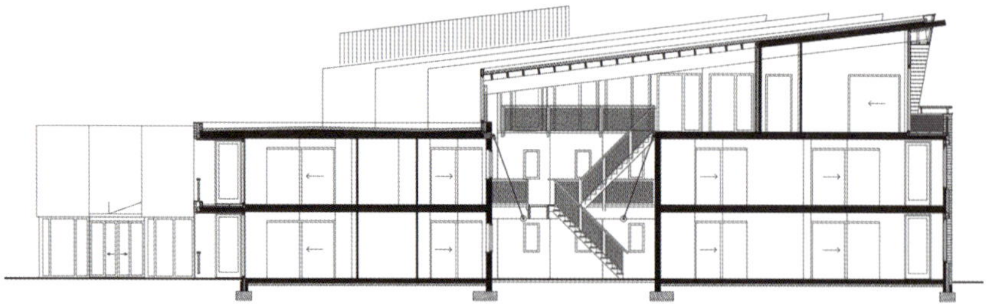

At the location of the former shopping centre "de Brink" and where a care home (Marishof) and ten sheltered houses were situated, has been designed a new care housing for elder people.

This projects exists in four parts: 30 houses for seniors (part A) orientated to the square "de Brink" and an extension of the Floralaan divided into two to three storeys, 54 residential homes for more elder people (part B) spread over three storeys situated at the Floralaan and Hagelkruis, 30 rooms for elder people who need more treatment and attention (part C). These parts are connected by the "meander" in this order (part D). The meander symbolises the phases of life which the future occupants will follow or experience, seen in time .

It is also the aorta of the complex that provides the following main functions: day care, recreation and welfare, treatment functions, management, administration and personnel, supporting services, space for nursery and care apartments.

The region and the history around Maarheeze (region Eindhoven) are a reason for the use of material, colour and texture. It is a translation of a historical rural substance into a coded message. This will be visible when intensively studying the building. In small amounts, images of "recognition" come up in the designers' consciousness and because of that, acceptance of relative large extra care housing in a small community is being simplified. The materials are distracted from the rural scenery and the process of transience. After a while they show light grey patina. The designers have developed a special concrete large brick for this project in the spirit of "Cradle to Cradle". It was produced not far from the site with low energy cost and it can be reused if necessary.

项目的原址是"边缘"购物中心、马丽莎夫养老院和10所庇护住宅，现在被改造成一个新的养老院。

项目由四部分组成：30套朝向"边缘"广场的老年住宅（A区）和弗拉兰扩建2到3层，位于弗拉兰和哈吉克鲁斯的54套老年公寓（B区），30套配备护理设施的老年公寓（C区），以及连接这些部分的走道（D区）。走道象征着人生的各个阶段，入住者将会在日后逐渐经历。

项目的主体部分提供了主要功能：日间护理、娱乐和福利设施、治疗功能、管理、行政和人事支持、辅助服务、托管和护理公寓。

马尔海济的区域特点和历史决定了材料、色彩和纹理的运用。这是一个将历史田园风格解码的过程，在对建筑的深入研究中得以体现。设计师十分注重形象的辨识度，因此，在小型社区内的大型养老设施必须要简化。材料的运用兼具田园风格，又经久耐用。一段时间后，它们会呈现出浅灰色做旧风格。设计师特别为项目开发了一种大型混凝土砖，采用了"从始至终"的绿色设计理念。这些砖块在不远处制作，低成本、低能耗，还可以被再利用。

1. The designers highlight the architecture's identity
2. The design emphasises the use of materials and textures
3. The building has a pastoral style
4. The design is simple and lively

1. 设计师强调建筑的形象辨识度
2. 设计注重材料和纹理的运用
3. 养老院十分具有田园风格
4. 简洁明快的房屋设计

Award:
International Brick Award 2008

获奖情况：
2008国际砖建筑奖

1

2

3

1

2

3

4

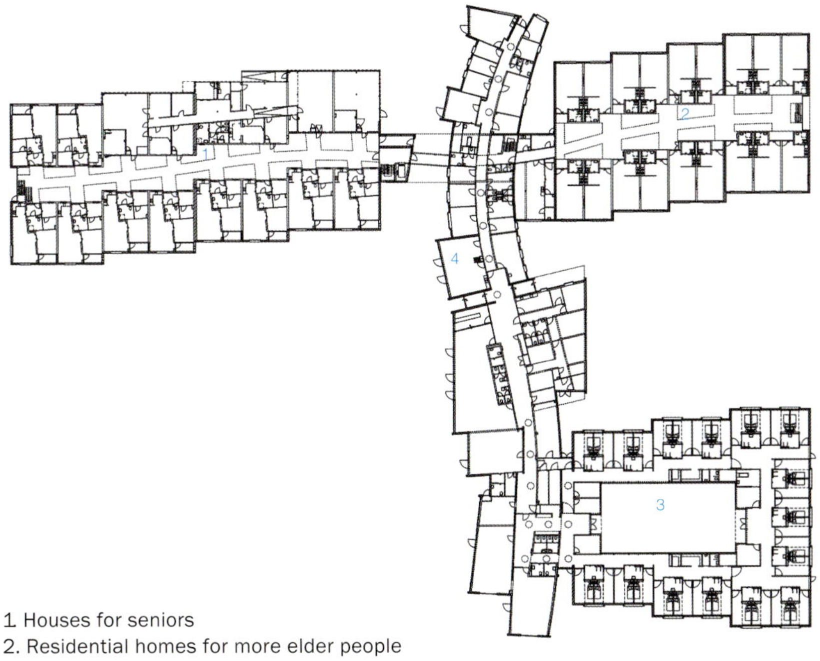

1. Houses for seniors
2. Residential homes for more elder people
3. Rooms for elder people who need more treatment and attention
4. Care units

1. 老年人房间
2. 高龄老人区
3. 治疗与关注区
4. 监护区

1-2. The delicate paving
3. French windows ensure the natural lighting of the interior space
4. The application of concrete and wood contrast with each other

1,2. 路面地砖铺装精巧
3. 大落地窗保证了室内的自然采光
4. 混凝土与木材的使用形成了对比

1

1. The main material used in the interior is wood
2. Wood textures are taken as the decoration
3. Ring patio
4. The inside natural landscape

1. 室内主要材料为木质结构
2. 设计采用木材的纹理作为装饰
3. 环形天井
4. 建筑内部的自然景观

穆里斯健康中心 Muros Health Centre

Location: A Coruña, Spain **Completion date:** 2006 **Designer:** Irisarri+Piñera **Photographer:** Juan Rodríguez **Area:** 2,400 m^2

项目地点：西班牙，拉科鲁尼亚 完成时间：2006年 设计师：Irisarri+Piñera设计公司 摄影师：胡安·罗德里格斯 面积：2,400平方米

The project tries to control the immediate environment by creating in its interior and its own landscape, since the existing surroundings it sits amongst have an erratic character, some of them exist with heights lower than the new planning restrictions permit and others are plots empty at the moment. For this the building generates a continuous courtyard north to south, its scale nuanced and enriched by the pieces that cross it, and in that the party walls are one more façade to superpose overlapping lattice of wood, allowing to qualify and to control it from the project itself. This way the building is conceived as an excavated block, leaving the extracted turned so much into exterior court, as into interior zones of wait.

The ventilated polycarbonate façade tints the relation with the exterior and the interior, leaving the presence of the air and the "green" in the abandoned courtyard of apples and perceptively increasing the dimension of the north-south courtyard, on having blurred its limits and having increased its light; it also is tremendously effective in terms of energy, and allows to control the various holes that the building demands, integrating them with in a unifying surface.

Three materials generate diverse atmospheres that are linked in the experience of the user, exploring the precise construction, working in the presence of objects in the landscape, and the control of the relationship between the constructed space and the environment.

The building envelope with a wood supporting structure, offers many nuances and an extraordinary interaction with environmental conditions. The glass of the exterior façade is allocated in two layers; it varies between being a veil, forming a ventilated façade that lets compacted wood shaving insulation be seen, a translucent but thick wall, wood between the two sheets of glass, a light lattice that opens the building to the exterior.

由于其所处的环境具有不稳定的特点，一些建筑过低，还有一些空间是空地，项目试图通过自身的室内和景观设计来控制周遭环境。建筑打造了一个南北向延伸的庭院，其规模随着建筑的构成而变化。界墙的外围是低价的木栅格，控制了项目的范围。这样一来，建筑形成了一个被挖空的块状结构，留下了外部庭院和内部等候空间。

通风聚碳酸酯外墙联系了室内外，突出了空气和废弃苹果园中绿色植物，也通过模糊了界限和增加亮度从视觉上增加了南北庭院的尺寸。外墙还十分节能，可以控制建筑所需的孔洞，并且使它们的外观统一。

三种材料营造出不同的氛围，这些与使用者的感受、对精确结构的探索、建筑与景观的关系以及建筑空间和环境的关系控制有紧密的联系。

建筑的木支撑结构打造了许多细微的差别，也与环境形成了非凡的互动。外墙分为两层，两层玻璃之间是压紧的木刨花绝缘层。墙壁透明而厚重，像一个轻盈的栅格，让建筑与外部环境结合在一起。

1. A long-range perspective
2. The façade is made of glazed curtain wall

1. 建筑远景
2. 建筑表皮采用玻璃幕墙

1

2

1

2

3

1. Glass and wood form grids
2. The building is surrounded by landscape
3. The façade is energy-effective

1. 玻璃和木材形成了栅格形状
2. 建筑处在广阔的自然景观之中
3. 建筑外墙十分节能

1. Entrance	1. 入口
2. Reception	2. 接待处
3. Clinic	3. 诊所
4. Centralisation facilities	4. 中央设施
5. Social work	5. 社区工作区
6. Polyvalent	6. 多功能区
7. Treatment room	7. 治疗室
8. Extractions	8. 提取区
9. Expected	9. 预期室
10. Radiology contral	10. 放射治疗区
11. Cabins	11. 房间
12. RX room	12. 接收室
13. Revealed	13. 诊断区
14. Physiotherapy	14. 物理治疗区
15. Costumes	15. 病患区

1. There is a courtyard in the centre of the building
2. The mid-air bridge
3. The open interior space

1. 建筑中央形成了外部庭院
2. 空中通道
3. 室内空间开阔

3

1

2

3

4

1. A colourful interior design
2. Wood pathway
3. The interior layout is simple and clean
4. The clean corridor

1. 色彩丰富的室内设计
2. 木质结构通道
3. 内部布局十分简洁
4. 整洁的走廊

普罗维登斯癌症中心 Providence Cancer Centre

Location: Portland, USA **Completion date:** 2008 **Designer:** ZGF Architects LLP **Photographer:** Eckert & Eckert **Area:** 45,708 m^2

项目地点：美国，波特兰 完成时间：2008年 设计师：ZGF建筑事务所 摄影师：埃克特&埃克特 面积：45,708平方米

The Providence Cancer Centre unites a variety of diagnostic, treatment, research and recovery services under one roof in a new 11-storey facility. This project is a component of the North Pavilion and new Northwest Facility at Providence Portland Medical Centre (PPMC), which also houses clinical and resource space for Providence Child Centre, a 900-car parking garage and an enlarged and upgraded Central Utility Plant.

The project offered an opportunity to create a new image and identity for PPMC. The North Pavilion is highly visible from the freeway and one of the first major buildings seen by travellers and commuters as they enter Portland. The massing and architectural expression of the project will reshape and enhance public perception of the campus and its programmes. Additionally, a new front door for the Cancer Centre was established on NE 47th Avenue, highlighting one of the region's premier research and treatment programmes.

The North Pavilion building plan was configured to fit the limited project site, to optimise functional connections to existing buildings and to maximise the dramatic views of Mt. Hood, Mt. St Helens and downtown from patient rooms on the tower's upper floors. The L-shaped tower plan orients the major building façades to the east and north to capture views. Between the tower and the main hospital, a central courtyard space provides access to light and views for existing patient rooms and establishes a new pedestrian path through the centre of the campus.

The design team wanted to produce a building that fit well within the existing campus and at the same time create a modern image for the new facility. Warm brick was chosen as the major exterior material to match other historic buildings on the campus. To add a feeling of science and technology to the building, metal panels, glass and sun screens were used to complete the exterior pallet.

The project design focuses on the patient. Nursing is decentralised, giving floors a residential, hospitality atmosphere; this also enhances nurses' ability to monitor patients, with all charting areas located adjacent to the rooms. All patient rooms are private rooms with large windows, and a residential look and feel to the furnishings, cabinetry and material palette. Patient rooms include a family zone, a space where family members can gather while the patient is receiving treatment, or spend the night on a pull-out sofa. Every room has not one, but two, flat screen televisions, to accommodate different interests. There are also family lounges located at the ends of patient floors as well as kitchenettes and other amenity spaces for families.

普罗维登斯癌症中心集诊断、治疗、研究和恢复服务于一身，是一座11层楼高的医疗机构。项目由北馆和位于普罗维登斯波特兰医疗中心的西北部设施组成。普罗维登斯波特兰医疗中心还包含普罗维登斯儿童中心的医疗和资料室、900个停车位的车库和扩建升级的中央事业设备。

项目为普罗维登斯波特兰医疗中心打造了全新的形象。从高速公路可以清晰地看到北馆，是人们进入波特兰后首先见到的主要建筑之一。项目的造型和建筑风格重塑，提高了园区的公共认知度。此外，癌症中心在NE47街上的新入口突出了该地区的主要研究治疗项目。

北馆的规划依据受限的场地而进行，优化了与原有建筑的功能连接，让高层病房可以最大限度地享受胡德山、圣海伦斯山和市中心的景色。L造型让建筑的主要部分朝向东侧和北侧，以获取最佳的风景。塔楼和主医院之间是一个中心庭院，为原有的病房提供了采光和风景，同时也在园区中心打造了一个新人行道。

设计团队试图打造一座与原有园区相适应，同时又拥有现代形象的新设施。具有温和感的砖块被选为外墙的主要材料，与园区内的历史建筑相映衬。为了增强建筑的科技感，外部结构还应用了金属面板、玻璃和遮阳板。

项目设计以患者为中心。护理被分散化了，让各个楼层拥有一种居家的热情氛围。而将图表制作区安排在病房旁也能提高护士对患者的监护能力。每间病房都拥有巨大的窗户和具有家居感的装饰、家具和材料。病房有一个家属区，在患者接受治疗时供家属聚在一起或是睡在折叠沙发上。每间房都有两台平板电视供人选择。病房楼层的尽头还设有家属休息室、小厨房和其他便利设施。

1. Aerial view of the architecture
2. A side view of the centre
3. The transparent glass entrance

1. 建筑俯瞰
2. 医疗中心侧面
3. 通透的玻璃入口

3

1

1. Trapezoid lighting fixtures in front of the centre
2. The design of central hall is quite fluid

1. 医疗中心前方的梯形灯饰
2. 中央大厅设计具有流线感

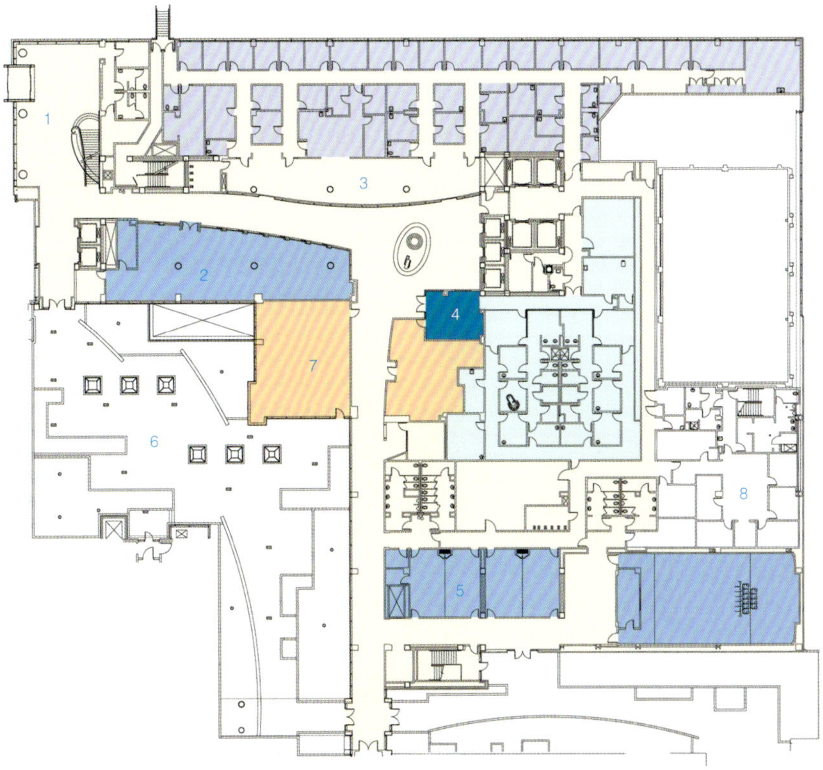

1. Entry lobby
2. Jill Lematta learning centre
3. Cancer support services
4. Spa
5. Auditorium
6. Garden
7. Café
8. Central utility plant

1. 入口大厅
2. 吉尔·拉玛塔学习中心
3. 癌症辅助服务
4. 水疗中心
5. 礼堂
6. 花园
7. 咖啡厅
8. 中央设备区

1

2

1. There is a small fountain in the atrium
2. The corridor with good lighting
3. The warm interior design
4. The interior layout is intimate
5. Reception

1. 中庭设有小喷水池
2. 采光良好的走廊
3. 温馨的室内设计
4. 室内布局让人产生亲切感
5. 接待台

3

4

5

1. The spacious and bright patient room
2. Waiting area
3. Medical treatment area

1. 明亮宽敞的病房
2. 休息区
3. 医疗工作区

3

斯特拉思登18套精神科病房 Stratheden 18-Bed Dementia and Mental Health Unit

Location: Stratheden, UK **Completion date:** 2009 **Designer:** Richard Murphy Architects **Photographer:** Richard Murphy, Graeme Armet, Brian Tobin, Tersius Maass **Area:** 3,432 m^2

项目地点：英国，斯特拉思登 完成时间：2009年 设计师：理查德・墨菲建筑事务所 摄影师：理查德・墨菲、格雷姆・阿尔梅特、布莱恩・托宾、特尔西斯・马斯 面积：3,432平方米

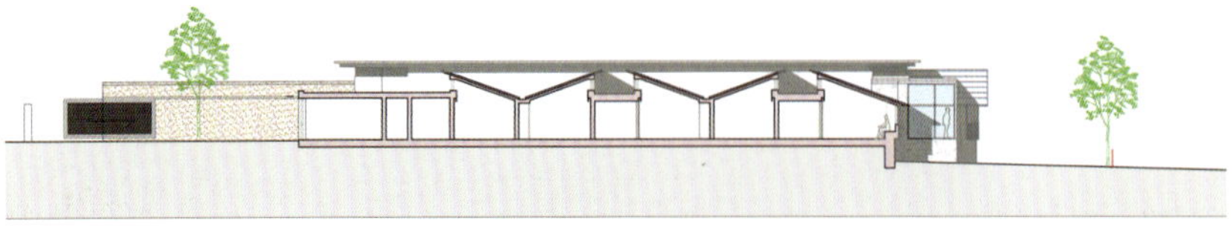

The proposal was to provide an 18-bed low security dementia unit in the grounds of NHS Fife Stratheden Hospital. The accommodation consists of 18 single bedrooms with en-suite shower rooms and associated accommodation both for patients and staff.

The building is a single storey unit with the bedrooms split into two wings, male and female, with communal facilities located centrally to the north to form a U-shape creating a south-facing secure garden for patients. The south edge of the garden is walled and frames the view south towards Walton Hill and White Hill. The design allows for patients to wander freely around the building and into the secure sensory garden.

Access to the building is via a new footpath. At the entrance is a waiting/relatives room with water closet and shower room. Beyond the double doors is the patients area with a centrally located nurses base. In the middle of the plan are the patient sitting and dining rooms connected by folding sliding doors and looking onto a south-facing veranda and the secure garden beyond. From these two rooms the two wings are clearly visible. To the left of the main entrance is the staff area containing staff facilities and consulting rooms, access to this area is restricted.

The two wings contain all the bedrooms and en-suite shower rooms; each bedroom has a bay window with views out into the grounds of Stratheden. Secondary light comes into each individual room via a roof light located below the pitched roof ridge line. Each bedroom is identifiedable by individual pitched roofs giving the patients a sense of their own identity within the building. For many patients this will be their last home. All the other accommodation and circulation areas have flat roofs at different heights. The wings have small alcoves with built-in seating for patients to stop and sit and look out into the garden.

项目是在斯特拉思登医院打造了18套低安全系数迟滞患者病房，由18套带浴室的独立病房和配套食宿设施组成。

单层建筑中的卧室被分为两半，分为男子病房和女子病房。公共设施位于中间，朝北，形成了一个U字形，在中间为患者打造了一个朝南的安全花园。花园的南侧建造了围墙，外面是沃尔顿山和白山的风景。设计让病人可以自由地在建筑周围和安全感官花园中漫步。

一条新的小径通往建筑。入口的等候/家属室带有洗手间和浴室。双开门后是病患区，中央是护士台。设计的中心是病患休息室和餐厅，与折叠拉门相连，面朝南的游廊和后面的安全花园。从这两个房间可以清晰地看到建筑的两翼。主入口的左侧是员工区，包含员工服务设施和资讯室，进入这一区域将受到严格的限制。

两翼包含全部的病房和配套的浴室，每间病房都有一个飘窗，可以遥看斯特拉思登的风景。斜屋顶屋脊下方的天窗为室内提供了第二照明。每间病房都有独立的斜屋顶，保证了患者有独立的存在感。对很多患者来说，这里将是他们最后的家。其他设施和流通区域则采用了高低不同的平屋顶。两翼设有带内置式座椅的壁龛，供患者随时坐下观看花园的风景。

1. A long-range perspective
2. The architecture is located in a peaceful natural environment
3. The design has a unique form

1. 建筑远景
2. 建筑处在宁静的自然环境中
3. 设计造型独特

Award:
RIBA Award Scotland Award 2011

获奖情况：
2011年英国皇家建筑师协会苏格兰奖

1

2

3

1

2

1. It is a graceful environment of the courtyard
2. Large windows ensure the interior lighting
3. Each room has a bay window

1. 中央庭院环境优美
2. 大型窗子保证了室内采光
3. 每间病房都设有一扇飘窗

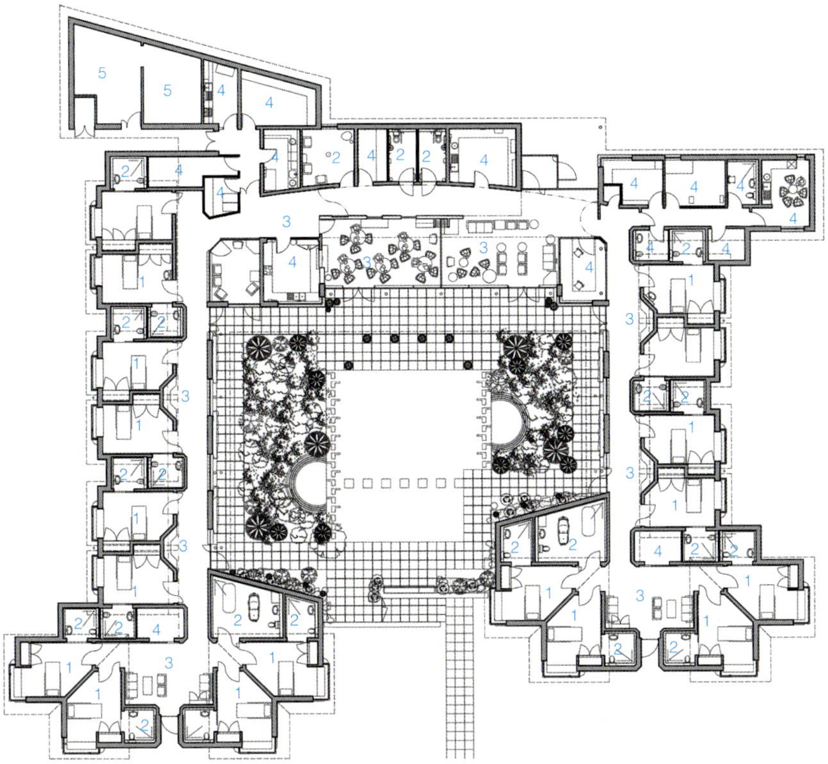

1. Bedrooms
2. Bathroom
3. Patient areas
4. Staff/consulting
5. Services

1. 卧室
2. 浴室
3. 病患活动区
4. 办公/诊疗室
5. 服务设施

1. The entrance
2. The corridor design is concise and generous
3. The corridor leads to different functional areas
4. Each room is equipped with individual pitched roofs

1. 入口处
2. 通道设计简洁大方
3. 走廊通往不同的功能区
4. 每间病房都配有斜屋顶

3

4

罗斯福温泉疗养院——布兰卡德大厅

Blanchard Hall – Roosevelt Warm Springs Institute for Rehabilitation

Location: Georgia, USA **Completion date:** 2008 **Designer:** Stanley Beaman & Sears **Photographer:** Jim Roof Creative **Area:** 3,250 m^2

项目地点：美国，乔治亚 完成时间：2008年 设计师：斯坦利·毕曼&希尔斯 摄影师：吉姆·卢夫创意公司 面积：3,250平方米

The Blanchard Hall Outpatient Centre was conceived not only to accommodate clinics, a rehabilitation gymnasium, and fabrication workshops for orthotics and prosthetics, but also to function as a new gateway to the institute, which was founded in 1927. Existing campus buildings, reflecting the classical revival aesthetic of their time, are of red brick with white-colonnaded porticos and connecting arcades. Blanchard Hall's design breaks dramatically with that style, while reinterpreting elements of its vocabulary in contemporary terms and materials.

Traditional exterior surfaces – stucco and brick in a similar colour to that of the existing buildings – transition into façades of glass and steel. A two-storey entry lobby refers to the similarly-scaled "great halls" of older campus buildings. Portico and colonnade are present, but reimagined in industrial-modern materials; these evoke the technology of today's prosthetics and orthotics, with piston-like steel rods and tensile fabric mesh canopies, which mimic the movable parts and stretchy carbon-fibre-and-fibreglass material typically used in fabricating prosthetics.

A difficult site required the structure to be set into a steep slope, and to refrain from damaging a creek which separates it from its parking area. The building's generous roof overhang combines with the hillside to create a sense of protection while focusing attention on its glass walls which convey an open welcome. The challenge of the creek was solved with a wheelchair accessible bridge that not only provides access but offers views into the two-storey therapy gym. From inside, the glass walls maximise views of natural environment and allow abundant natural light.

With its prominent site on the campus' main drive and its exuberant design, Blanchard Hall reflects the progressive approach to care in rehabilitation medicine practiced at the Roosevelt Warm Springs Institute.

布兰卡德大厅门诊中心不仅包含门诊部、康复健身中心、矫形和假肢装配工作室，还是通往建于1927年的疗养院的门户。原有的建筑反映了当时的经典美学价值，采用了红砖、白色列柱门廊和拱形游廊。布兰卡德大厅的设计颠覆了这一风格，以现代元素对建筑进行了全新的诠释。

与原有建筑色彩类似的灰泥和砖块等传统材料被换成了玻璃和钢铁。两层高的入口大厅与旧建筑的大礼堂类似。列柱和门廊得以呈现，但是采用了现代工业化材料。活塞钢条和拉伸纤维网华盖模仿了假肢装配中的可移动部分和拉伸纤维玻璃材料，让人想起了当今的假肢和矫形技术。

场地要求建筑与陡坡相适应，并且不能破坏隔开它和停车场之间的小溪。建筑宽大的屋顶悬挑出来，与山坡结为一体，形成了一种防卫感，而玻璃幕墙则开放而热情。小溪的问题通过无障碍小桥得以解决，小桥不仅起到了通道作用，还为二楼的理疗健身中心提供了风景。玻璃幕墙将室内的可视自然景观最大化，还提供了充足的自然光。

大厅位于园区的主要机动车道上，设计十分抢眼，反映了罗斯福温泉疗养院在康复医疗实践中的积极处理方式。

1. The connecting arcades in front of the building
2. The new portico uses modern industrial materials
3. Glass and steel replace the traditional bricks

1. 建筑前的白色游廊
2. 新建的门廊采用现代化工业材料
3. 玻璃和钢铁材料代替了传统的砖瓦

1

2

3

1

2

1. The building's generous roof overhangs
2. A side view of the clinic
3. The colonnades of steel
4. At night, the portico is full of modern sense

1. 建筑宽大的屋顶悬挑出来
2. 医院侧面
3. 钢铁材料建造的列柱
4. 夜色中的门廊充满现代气息

3

4

1. Main lobby & reception　1. 主大厅/接待处
2. Waiting　2. 等候区
3. Therapy gym　3. 理疗健身室
4. Pediatric therapy gym　4. 儿科理疗健身室
5. Treatment room　5. 治疗室
6. Speech therapy　6. 言语治疗室
7. Business office　7. 办公室
8. Exam room　8. 检查室
9. Conference room　9. 会议室

1

1. The interior design is elegant and calm
2. Rehabilitation area
3. Nurse Station

1. 室内设计高雅沉静
2. 运动康复区
3. 护理站

杜祖斯岛护理医院 Hospice Djursland

Location: Djursland, Denmark **Completion date:** 2007 **Designer:** C. F. Møller Architects **Photographer:** Adam Moerk **Area:** 1,900 m^2

项目地点：丹麦，杜祖斯岛 完成时间：2007年 设计师：C・F・穆勒建筑事务所 摄影师：亚当・默克 面积：1,900平方米

The Hospice Djursland is a palliative treatment facility with room for 15 patients, located in a beautiful landscape setting overlooking the bay of Aarhus. In hospice design, the architect's finest task is to create surroundings which will provide the best possible conditions to promote quality of life, respect and a dignified death.

Djursland Hospice is first and foremost a building within a landscape. No matter where you go in the building – the reception area, the garden of the senses, the atriums, the staff room, the lounge, the reflection room or the patient rooms – the beautiful landscape is always present.

The architects have aimed to create a very humane building; by which they mean a building which is not an institution, but rather a home which provides adequate physical and mental space for those who will live there in their final time, as well as for their relatives and the staff.

The semi-circular layout is to ensure that all patient rooms enjoy an equally privileged view over the bay, and that they are located within a more private zone in the building, set back from the common rooms. Each room has a private terrace overlooking the landscape, and the section of the roof draws daylight deep into the rooms, providing a skylight over the sleeping area and the bathroom, and a soft curve to the ceiling.

The common materials used throughout are copper, oak and glass, which interact beautifully and naturally with the landscape and provide a sense of warmth in the rooms.

The landscape branch of C. F. Møller Architects has designed the landscape and gardens surrounding the hospice, with special emphasis on the sensory aspects of sight, smell, touch and sound, as well as the overall accessibility for patients, even those confined to beds, resulting in a series of soft, rounded shapes and niches joined by green rubber-bitumen surfaces.

杜祖斯岛护理医院是一家舒减疗法医疗设施，共有15套病房，位于风景优美的杜祖斯岛，可以远眺奥尔胡斯湾。护理医院的设计中，建筑师的首要任务是打造最佳的环境，为病患提高生活质量，让他们度过有尊严的余生。

首先，杜祖斯岛护理医院是一座风景中的建筑。在医院的各个角落，无论是接待处、感官花园、中庭、员工工作室、休息室、沉思室，还是病房，随处可见优美的风景。

建筑师力求打造一座人性化的建筑，不是建造一座医疗设施，而是一个家。病患将在这里度过自己最后的旅程，而医院将为他们和亲属以及工作人员提供舒适的生理和心理环境。

半圆形平面布局让所有的病房都能享受海湾的美景，也让它们处在一个相对私密的区域，远离公共空间。每间病房都有一个私人露台，屋顶的切面让日光照进房间的深处，为卧室和浴室提供了天窗，也在天花板上划出一道柔和的弧线。

建筑广泛使用了铜、橡木和玻璃作为建筑材料，它们与美丽的自然景观形成了互动，也为室内增添了暖意。

C・F・穆勒建筑事务所的景观团队对医院周围的景观和花园进行了设计，特别注重了视觉、嗅觉、触觉、听觉等感官体验和病患是否容易接近。就连花坛中的景观都采用了柔和、圆滑的造型和橡胶沥青表面。

1. The main volume of the architecture
2. The semi-circular layout ensures all patient rooms to enjoy a view over the bay

1. 建筑主体造型
2. 半圆形平面布局让每个房间都能欣赏到海湾的美景

1

2

1. The architecture applies copper, oak and glass as main materials
2. The French windows in the patient room ensure great daylight
3. A side view of the circle building

1. 建筑采用铜、橡木和玻璃作为主要材料
2. 病房建有落地窗保证采光良好
3. 环形建筑侧面

1. Entrance	1. 入口
2. Foyer/café	2. 大厅/咖啡厅
3. Guest accommodation	3. 来宾病房
4. Meeting	4. 会议室
5. Assembly hall	5. 集会大厅
6. Meditation	6. 冥想室
7. Accommodation	7. 病房
8. Bathroom	8. 浴室
9. Administration/offices	9. 行政办公室
10. Courtyard gardens	10 .庭院花园
11. Music room	11. 音乐室
12. Lounge/fireplace	12. 休息室/壁炉
13. Library	13. 图书室
14. Public lounge/play area	14. 公共休息室/游戏区
15. Treatment rooms	15. 治疗室
16. Spa rooms	16. 水疗室
17. Kitchen	17. 厨房
18. Garage/workshop	18. 车库/工作室
19. Reflecting pool	19. 倒影池
20. Sensory gardens	20. 感官花园
21. Playground	21. 操场

3

1

2

3

4

1. Patients can walk freely inside the building
2. The ring-shaped corridor leads to different areas
3. The patient room
4. The patient room design has a home-like atmosphere
5. The room has a great view

1. 患者可在建筑内部空间自由随意的活动
2. 环状走廊通往各个区域
3. 病房布局
4. 如家一般亲切的病房设计
5. 病房视野良好

5

台湾大学医学院附设医院儿童医疗大楼

Children's Medical Centre of Taiwan University Hospital

Location: Taipei, China **Completion date:** 2008 **Designer:** C.C.Hsu & Associates of Architects and Engineers; NBBJ **Photographer:** Zheng Jinming **Area:** 73,870 m^2

项目地点：中国，台北 完成时间：2008年 设计师：许常吉建筑师事务所；NBBJ 摄影师：郑锦铭 面积：73,870平方米

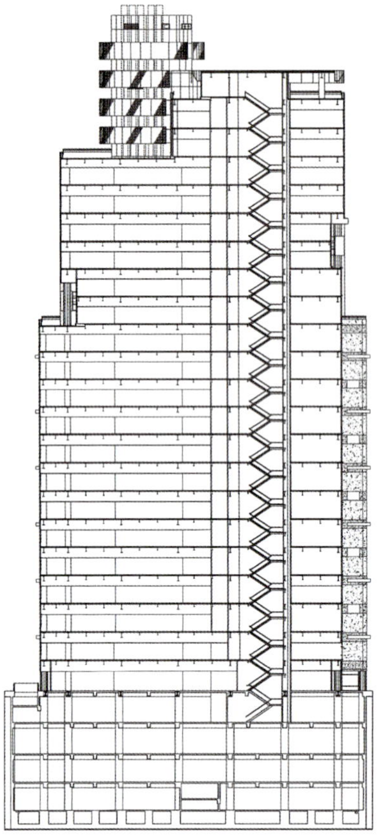

This is the City in the Sky, the representative of hope, the cohesion of imagination. In the concept of "Children are not the miniatures of adult, nor little men. Their values and potentials are priceless as diamonds", and taking over 20 years work, the first children hospital financed by government is built in 2008.

The architecture is comprised with four coloured volumes. The square and banding volume with openings is the main component. The combination of precast tiling concrete and rhythmic openings takes the convenience of construction into consideration. The square volume with openings is the highest one. The heavy and strong façade highlights the verticality of the architecture. The other interlaced volume, in contrast, adds some interesting horizontal elements. The glazed façade volume clearly defines the important point in the northeast corner. The energy-efficient façade will become a remarkable lantern in the night, just like a lighthouse of hope and direction.

The interior is rich in colours, reflecting children's pure and lively features. The goal is to create an intimate environment for the children, and build their bonds with the hospital. The interesting forms, colour tones and tiling make the architecture full of feature, energy and inspiration. The smart integration with the public art is another feature. The lighting and space plan are all specially designed for the children. For example, the child CT room is made into a capsule to reduce the fear of the young patients.

这是一座天空之城，是希望的代表、想象力的凝聚。秉持“小儿非大人的缩影（miniature），也不是小型的大人（little man），宛如钻石般，其价值与潜力无可估计”的认知，历时二十余年擘画，台湾第一所由政府资助建立的儿童医院正式于2008年落成启用。

建筑造型由4个主要的色块体量组成。方形和带状开窗体量是建筑构成的主体，贴砖混凝土预铸板结合韵律的开窗，考虑了施工模矩的方便性。方形开窗体量是最高的体量，厚实的墙面强化了建筑垂直的厚重感，另一交错相嵌的体量则以不同的带状风格，轻快地加添了建筑水平向的趣味性。玻璃帷幕体量清楚界定了东北角的重要端点，具节能效果的低反射玻璃面，在白天则产生相对于主体量的虚实趣味，夜晚则跃身为一显著的大灯笼，宛如一座象征希望和方向的灯塔。

在颜色的选择上，内部设计色彩丰富，以反映出儿童纯真、活泼的特质为主，造型极富童趣。目的让儿童产生对环境的亲切感，有助于在小小心灵里创造"属于我们的医院"的认同感。童趣的造型、色调及贴砖变化，使建筑独具特色，充满活力和启发性。与公共艺术巧妙进行结合，亦为本建筑一大特色。不论在灯光、空间规划等各方面，均特别为儿童着想，甚而将儿童断层扫描室改造为太空舱，以降低小病患恐惧感等贴心设计。

1. The architecture is comprised with four coloured volumes
2. A side view
3. The building consists of four main colour masses

1. 医院大楼外部色彩丰富
2. 侧面投影
3. 建筑造型由4个主要的色块体量组成

1

2

3

1

2

3

1. The interior design is bright
2. Playful forms
3. The red pendant adds vividness to the interior space

1. 医院内部设计明快
2. 设施造型极富童趣
3. 红色吊灯的设计增添了室内的活泼感

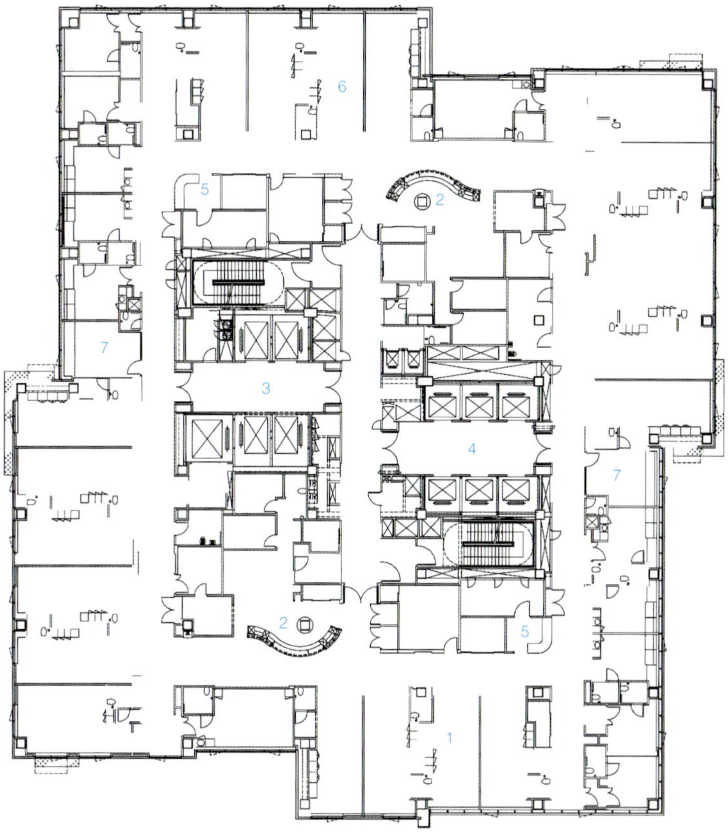

1. Patient rooms
2. Nursing station
3. Bed elevator
4. Visitor elevator
5. Sub nursing station
6. Patient rooms
7. Game and sunlight room

1. 病房区
2. 护理站
3. 移动病床电梯厅
4. 访客电梯厅
5. 次护理站
6. 病房区
7. 游戏兼日光室

1. Playroom in the physiological examination area
2. Waiting area in the outpatient

1. 生理检查室之儿童游戏间
2. 门诊等候区

2

1

1. The design creates a intimate environment for the children
2. The interior mouldings and colours are quite special
3. The wall is light and full of interest

1. 充满童趣的设计让儿童产生对环境的亲切感
2. 室内造型与色彩充满特点
3. 墙体设计轻快，富有趣味性

2

3

莱加西儿童医疗中心 Children's Medical Centre at Legacy

Location: Plano, Texas, USA **Completion date:** 2008 **Designer:** Doss Mabe, ZGF Architects LLP; Josh Theodore, PageSoutherlandPage **Photographer:** Robert Caneld, Pete Eckert **Area:** 29,730 m^2

项目地点：美国，德克萨斯州，普莱诺 完成时间：2008年 设计师：道斯·马贝，ZGF建筑事务所；约瑟·赛奥多，PSP设计事务所 摄影师：罗伯特·甘菲尔德、皮特·埃克特 面积：29,730平方米

Sited on an 84-acre suburban Plano Texas site, Children's Medical Centre at Legacy was conceived of as both a compliment to the main campus of Children's Medical Centre of Dallas, and as a community-based hospital capable of serving the system's low-acuity clientele closer to their north Dallas homes. Designed for future flexibility, the facility currently features three above-ground floors, plus a lower garden level, with 72 single-patient rooms, and four operating rooms. An additional eight operating rooms have been shelled (for a total of 12). Initial construction also includes full-service diagnostics with CT and MRI, an urgent/emergency care centre, a lobby café, full-service cafeteria, chapel, gift shop, family resource centre, and various patient and family lounges. Long-range plans allow for the addition of two floors atop the bed tower (with a potential expansion from 72 beds over three floors to 120 beds over five floors) and, if needed, the construction and connection of an additional five-level, two-wing bed tower that would add another 120 beds (for a total of 240).

Children's Medical Centre at Legacy shares its site, an undulating, grass-covered parcel bisected by a tree-lined stream, with a new ambulatory care facility and central plant. From its semi-circular glass entry façade, the building is open and inviting. A slant-roofed cylinder, clad in colourful blue-green aluminium shingles, rises beyond the façade and is juxtaposed against angular Texas limestone walls. The two patient wings extend east from the building's curved circulation spine, forming a courtyard in between. The overall look and feel of the new Children's Medical Centre at Legacy is that of an inviting community centre rather than an institutional medical facility.

Seeking to be child-right but not childish, the design of Children's Medical Centre at Legacy was heavily influenced by the Medical Centre's mission of providing patient-and family-centred care and using nature for its impact on healing. Ideas and feedback to help guide design decisions for the new facility came from a design advisory panel that consisted of patients, their families, hospital staff and caregivers, as well as from research compiled by the Centre for Health Systems and Design at Texas A&M University. From every public space to every patient room and corridor, the design team considered how patient care and safety issues could be enhanced by nature, daylight, and colour, while positively affecting the experiences of patients, families, and staff. What emerged was a set of guiding principles that were used as a standard to evaluate all aspects of the design. Throughout the process the team referenced studies regarding the variables of aesthetic qualities, along with productivity, quality of care, and safety in relation to the physical setting.

莱加西儿童医疗中心位于德克萨斯州普莱诺郊区一块339,935平方米的场地上，是达拉斯儿童医疗中心的主园区补充部分，为达拉斯北部的家庭提供基本医疗服务。医疗中心地上3层，地下1层，共有72间独立病房和4间手术室，另配有8间备用手术室（共计12间）。初期工程包括全套CT和核磁共振诊断设备、一个急救中心、一间大堂咖啡厅、服务齐全的食堂、礼拜堂、礼品店、家庭资料中心、以及各种患者和家属休息室。长期规划计划增添2层病房（将病床从72拓展为120）。如有需要另增5层和2幢翼楼，可再增加120张病床（共计240个床位）。莱加西儿童医疗中心与流动护理所和中心工厂共享绿草如茵、溪水环绕的场地。半圆形的玻璃入口让医院看起来宽敞而舒适。入口后面是圆柱体建筑，斜屋顶上覆盖着蓝绿铝板瓦，与生硬的得克萨斯石灰岩墙壁形成了鲜明对比。两个翼楼从建筑弯曲的走道向东延伸，在中间形成了一个庭院。莱加西儿童医疗中心的整体视觉效果让它看起来更像一个社区中心，而不是冷冰冰的医院。

1. The semi-circle glass entrance is open and bright
2. The limestone wall is quite special
3. The building has a unique form

1. 医院半圆形的玻璃入口开阔明亮
2. 石灰岩墙壁充满特色
3. 建筑造型丰富独特

1

2

3

1

2

医疗中心的设计以适合儿童为原则，却丝毫不幼稚，一切都以为病患和家属提供优质医疗服务为中心。一支由病患、家属、医院员工和护理人员组成的设计顾问小组为设计提供了观点和意见反馈，得克萨斯A&M大学健康系统和设计中心也提供了研究支持。从每一个公共空间到病房、走廊，设计团队对患者护理和安全实施进行了全面考虑，自然、采光、色彩都要对患者、家属和员工起到积极的影响。有一整套的指导原则来评估设计的各个方面。在设计过程中，设计团队对美观性、生产效率、护理质量和安全性进行了全方位的考虑。

1. The main lobby is bright and welcoming
2. Café is adjacent to colourful lobby

1. 大堂明亮，洋溢着热情的氛围
2. 咖啡厅设在彩色大堂的旁边

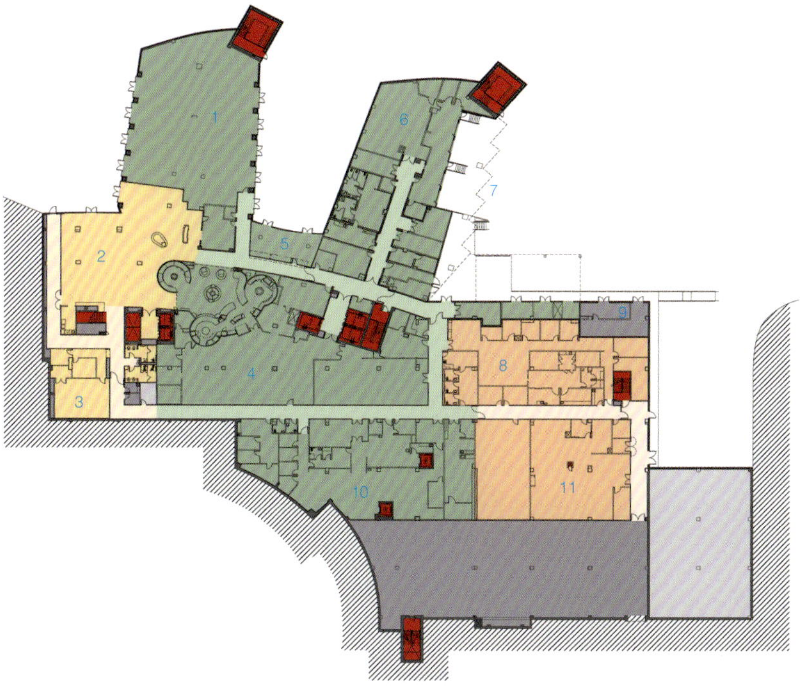

1. Education 1. 教育区
2. Dining 2. 餐厅
3. Chapel 3. 礼拜堂
4. Food services 4. 食品服务区
5. Physician dining 5. 医生餐厅
6. Materials management 6. 物资管理
7. Loading dock 7. 装卸区
8. Pharmacy 8. 药房
9. Mechanical 9. 机械室
10. Central sterile 10.无菌中心
11. Laboratory 11. 实验室

1. A quiet seating area on second floor above lobby
2. Public waiting room
3. A corridor links the hospital to a neighbouring medical office building

1. 大堂楼上安静的休息区
2. 公共等候区
3. 走廊将医院与隔壁的医疗办公楼连接起来

1. Family lounge with kitchen at end of each patient wing
2. Patient room
3. Large windows in rooms let in light and offer views to nature

1. 家属休息室设有厨房，位于病房楼的两端
2. 病房
3. 屋子里的大窗户提供了日光和优美的自然风景

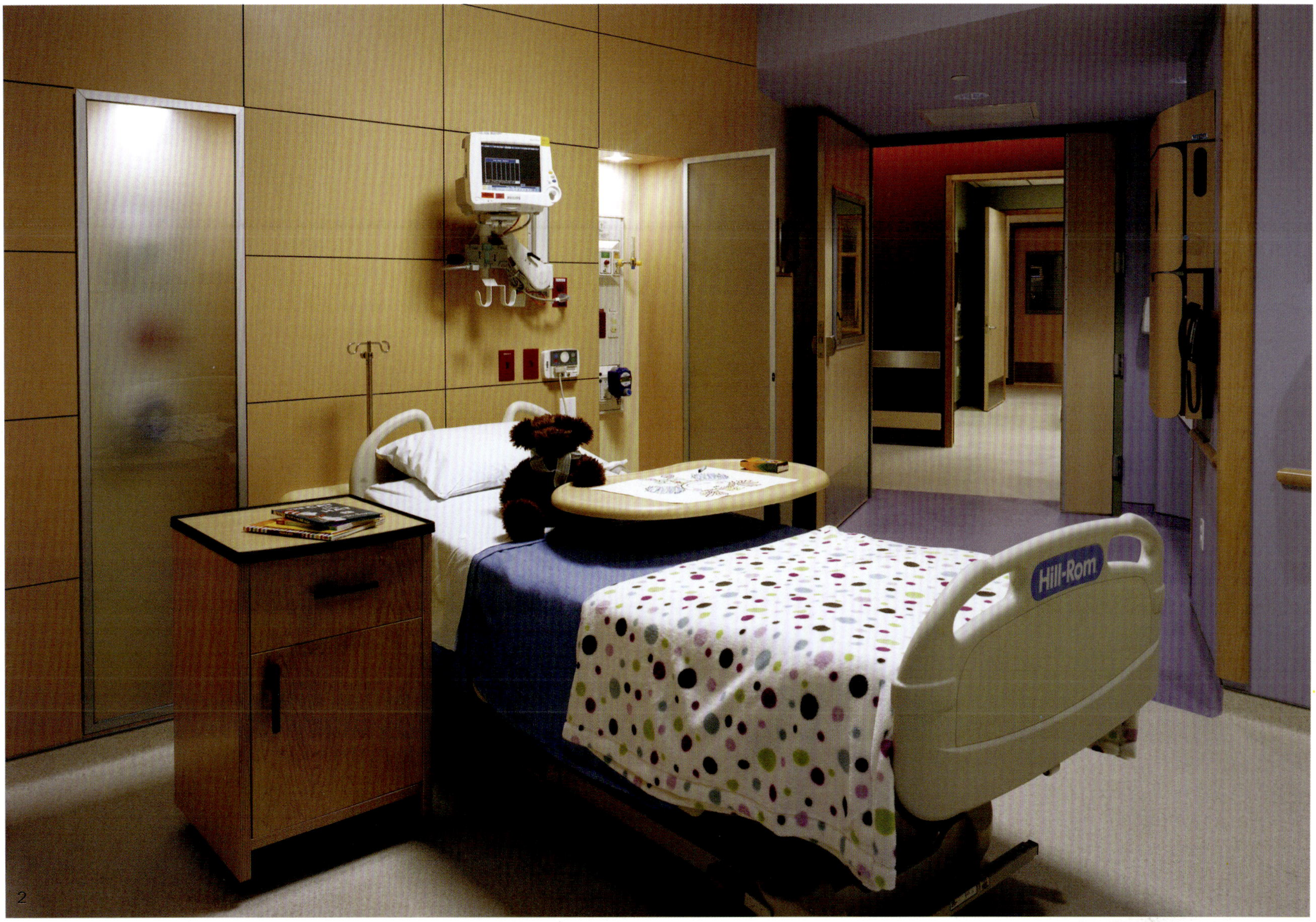

2

3

丹佛儿童医院 Denver Children's Hospital

Location: Aurora, USA **Completion date:** 2007 **Designer:** ZGF ARCHITECTS LLP **Photographer:** Eckert & Eckert, Basil Childers **Area:** 133,776 m^2

项目地点：美国，奥罗拉 完成时间：2007年 设计师：ZGF建筑事务所 摄影师：埃克特&埃克特、巴兹尔・切尔德斯 面积：133,776平方米

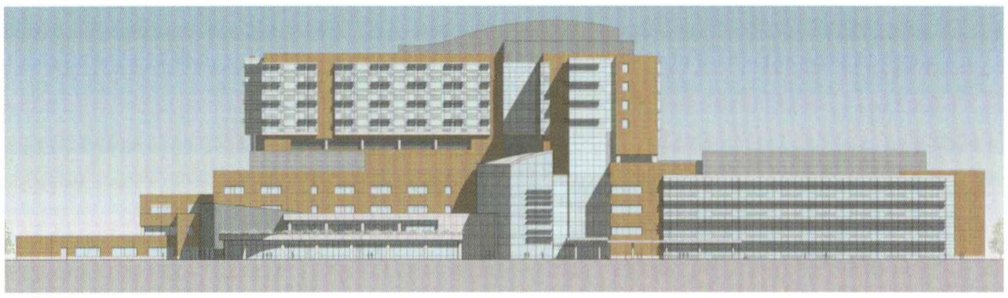

Situated on The University of Colorado, Denver Anschutz Medical Campus, the new Children's Hospital houses 270 patient beds, outpatient clinics, a pavilion for specialised services, offices, research and education space, and parking for 2,144 vehicles. Founded in 1908, The Children's Hospital (TCH) has consistently been ranked among the top 10 children's hospitals in the country by U.S. News & World Report.

The building itself is comprised of four programmatic components. These programmes include logistic services, diagnostic/treatment programmes, intensive patient care units and the patient care towers. The Lower Level of TCH houses logistic services and Level 1 accommodates diagnostic services and public spaces relating to the building entrance/lobby. Level 2 has programme elements that support surgical services and a blood donor centre. Level 3 includes the cardiovascular operating rooms and clinics, laboratory, CICU and PICU programmes and an exercise centre. Level 4 is the NICU floor and also contains family amenities, such as a family lounge, family laundry and family sleep rooms. Level 5 is a mechanical floor containing stacked air-handler units. Levels 6 are the "L" shaped patient care towers. Level 7 houses hematology and oncology services with a BMT suite.

Critical to the mission of Children's are the concepts of family centred care and healing. Specifically, the Children's Hospital had to revolve around patients and their families, providing a "most healing" environment for patients and a warm and inviting environment for families, friends and hospital staff. Designing such an environment, one embracing of family-centred care, led to a complex but rewarding evidence-based design process – a process that resulted in a hospital that expresses the depth and level of care that TCH is recognised for in both a medical sense and in the human sense.

丹佛儿童医院位于科罗拉多大学丹佛・安修斯医疗园区，拥有270个床位、门诊部、特殊服务馆、办公室、研究教育空间和2,144个车位的停车场。儿童医院创立于1908年，名列美国新闻和世界报道评选的美国十佳儿童医院。

建筑由3个功能区组成，分别为：后勤服务、诊疗项目、集中病患护理病房和患者楼。儿童医院的下层是后勤服务设施，1楼是诊疗服务和连接大楼入口大厅的公共空间。2楼是外科手术区和献血中心。3楼是心血管手术室和诊疗室、实验室、心脏重症监护室、儿科重症监护室和运动中心。4楼是新生儿重症监护室和家属便利设施，如家属休息室、洗衣房和卧室。5楼是机械设备区，放置着空调设施。6楼是L形的患者楼。7楼是血液肿瘤科，设有骨髓移植套间。

儿童医院的主要任务是打造以家庭为中心的护理和治疗服务。医院以患者和他们的家属为中心，为病人提供最具治愈感的环境，为家属、友人和医院员工提供最热情舒适的环境。以此为目标打造以家庭为中心的护理让设计过程变得复杂而有价值。设计让儿童医院兼具医疗特性和人性尺度。

1. The main volume is bright and clear
2. Glass is used as the main material for the façade
3. The stream-lined architecture

1. 医院主体通透明亮
2. 玻璃作为外墙主要材料
3. 流线型的建筑构造

1

2

3

1

2

3

4

5

1. A long-range perspective
2. Details of the structure
3. The hospital uses a large number of French windows
4. The interior space is open and generous
5. The design is playful

1. 建筑远景
2. 细部构造
3. 医院采用大量落地窗户
4. 内部空间宽敞广阔
5. 设计细节充满童趣

1. 24/7 fresh market place of café
2. Boettcher atrium
3. Chapel and garden of hope
4. Creative play
5. Emergency department
6. Family health library
7. Financial counseling
8. Gift shop
9. Main hospital entrance
10. Outpatient laboratory
11. Outpatient pavilion
12. Outpatient pharmacy
13. Patient access
14. Procedure centre
15. Radiology
16. Staff lounge
17. Volunteers and patients reception

1. 24小时咖啡厅
2. 博切尔中庭
3. 礼拜堂和希望花园
4. 创意游戏区
5. 急诊部
6. 家庭健康图书室
7. 财务咨询室
8. 礼品商店
9. 医院主入口
10. 门诊实验室
11. 门诊区
12. 门诊药房
13. 病患入口
14. 程序中心
15. 放射科
16. 员工休息室
17. 志愿者和病患接待处

1

2

1. The warm colours in the corridor add intimacy to the space
2. A corner inside
3. The public working area
4. The patient room is comfortable and warm

1. 走廊空间采用暖色调，增添亲切感
2. 室内一角
3. 公共工作区
4. 病房设计温馨舒适

3

4

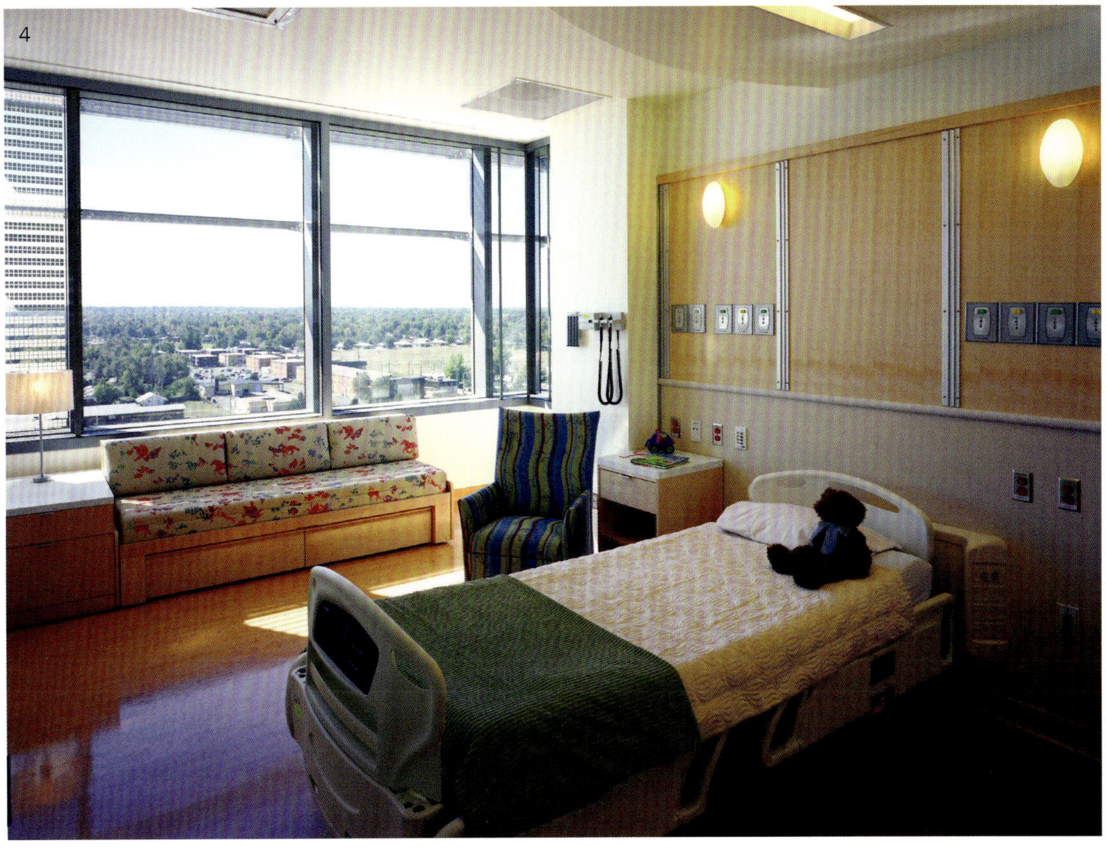

1

2

1. It is a lively and casual resting area
2. The clean interior design eases the hospital's depression
3. The spacious public dining area

1. 休息区设计活泼随意
2. 室内大方简洁的装饰减轻了医院的压抑感
3. 宽敞的公共就餐区

克芒儿童医院（妇婴门诊）

Kemang Children Hospital (Women and Children Clinic)

Location: Jakarta, Indonesia **Completion date:** 2008 **Designer:** ABODAY **Photographer:** Happy Lim Photography **Area:** 6,500 m^2

项目地点：印度尼西亚，雅加达 完成时间：2008年 设计师：ABODAY设计公司 摄影师：快乐的林摄影公司 面积：6,500平方米

The original land size of 2,600 square metres is barely enough to build a women and children clinic with 52 wards. The building programme is complex, which includes one floor dedicated for in vitro treatment. The five-storey building sits on an elongated site. Strict guideline of building setback results in a curvilinear building so thin that the part facing the road only has seven metres of width.

In a country where design is still not of paramount importance, the challenge is bigger when architect has to convince client with regards to their design vision that has nothing to do with client's commercial calculation. The budget is very modest for a clinic. Understandably client will invest more towards medical equipments rather than the architecture. Architect tries to marry the idea of having an excellent building that in return will help to boost the business. For a clinic that is intended to cater such a niche market, it is important to gain recognition from public using the uniqueness of design. As the client doesn't want to have a clinic with conventional approach, architect creates building façades that engage with the surroundings by using interwoven pattern of plaster concrete panel and colour glass. This is to mimic the surrounding area dominated by non-descript commercial building with colourful commercial signboard, at the same time enhancing the quality of clinic's interior by producing various colour glows for every room.

The idea of coloured glass façade came up upon reading a text about chromotherapy. It is part of natural healing that uses colour to balance energy wherever a person's body be lacking, be it physical, emotional, spiritual, or mental. Even though this has not been proven scientifically, the idea of using colour to affect people emotion is rather new and interesting. This seemingly irregular composition of glass and window on the façade is a result of room organisation within. Each room and facilities arranged along the main corridor that act as a spine of this curvilinear form. On the 3rd & 4th floor, these two-level corridors are connected by an irregular void in the wards waiting area, creating a fluid space called "the womb".

However complicated the façade seems, internal circulation and room programme is straight forward and clear. A long central corridor of 2.4 metres on each floor connects rooms and facilities distributed equally on both sides of the line. On each floor, this corridor centres on the reception lobby, which becomes a point of orientation for visitors. By dividing the building facilities into two linear blocks, each room is quite narrow in dimension, consequently resulting in an entire floor bathed in abundant natural light almost all day long.

原有2,600平方米的场地几乎不够建筑一座拥有52病房的妇婴门诊。建筑的规划十分复杂，其中的一层要用于试管疗法。五层楼高的建筑坐落在狭长的场地上。严格的建筑后移设计方案让医院呈弧形，临街的一面仅有7米宽。

在一个设计还未达到顶峰的国家，设计师必须尽力劝说委托人考虑与商业实际无关的设计理念。项目的预算十分有限，委托人希望将更多的资金投入到医疗设备，而不是建筑结构上。建筑师试图将优秀的建筑与推动商业发展结合起来。对于市场如此细化的诊所来说，通过独特的设计获得公众的注意力十分重要。由于委托人不想要一座传统的医疗建筑，建筑师通过石膏混凝土板和彩色玻璃的交织装饰了建筑的外墙。这与周围环境中五颜六色的布告板遥相呼应，同时也提升了诊所室内设计的质量，让每间房都能有彩色的光。

彩色玻璃外墙的理念来自于光谱疗法。那是一种自然疗法，通过色彩来平衡人体中缺乏的能量，如体质、情绪、精神或脑力。尽管这种疗法并没有被科学所证实，色彩能够影响人的情绪这一概念还是新鲜而有趣的。看似无序的玻璃和窗户排列方式由内部的房间设置决定。房间和设施沿着主走廊设置，在四楼和五楼的走廊由一个不规则的病房等候区连接，打造了一个流畅的空间。

尽管外墙设计十分复杂，内部循环路径和空间的设置简洁而直接。2.4米宽的中心走廊连接了位于两侧的房间和设施。每层楼走廊的中心都是接待大厅，是来访者的定向中心。由于线性设计，每间房间都十分狭窄，整层楼都能享用充足的自然光。

1. The building uses numerous stained glass
2. A long-range side view
3. The white façade matched well with the coloured glass

1. 建筑使用了大量彩色玻璃
2. 侧面远景
3. 白色外墙搭配彩色玻璃明亮活泼

1

2

3

1

2

3

1. The architecture highlights fluency
2. The side view of the building
3. The interior space is open and bright

1. 医院外型设计流畅
2. 建筑侧立面
3. 室内空间开阔明亮

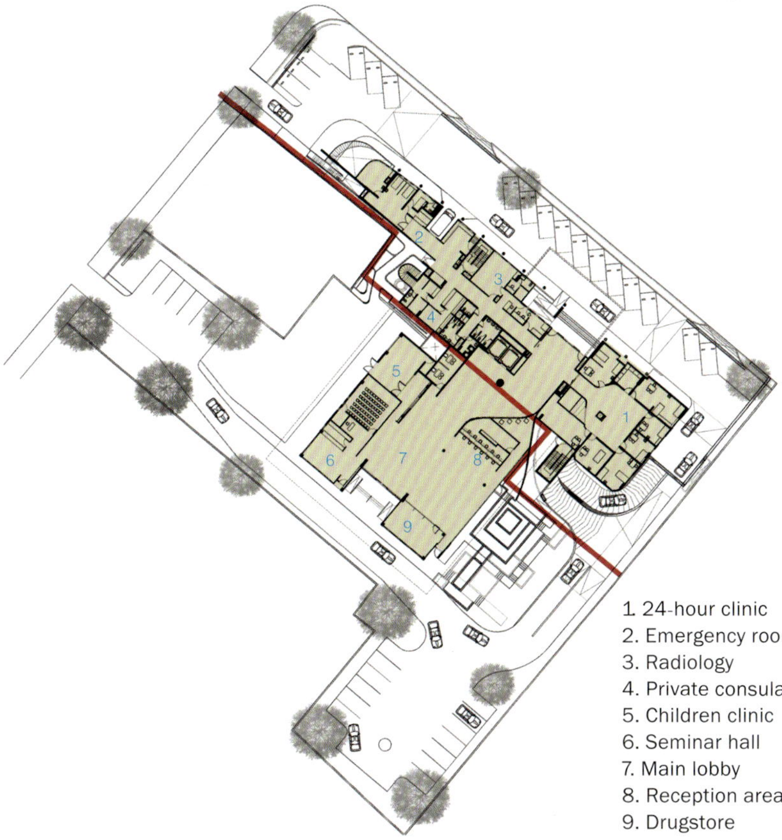

1. 24-hour clinic	1. 24小时门诊
2. Emergency room	2. 急诊室
3. Radiology	3. 放射科
4. Private consulation	4. 私人咨询
5. Children clinic	5. 儿科门诊
6. Seminar hall	6. 研讨大厅
7. Main lobby	7. 主大堂
8. Reception area	8. 接待处
9. Drugstore	9. 药房

1

1. The reception hall has a clean design
2. The comfortable and lively interior design
3. A comfortable patient room

1. 接待大厅设计简洁
2. 室内布置温馨活泼
3. 舒适的病房

皇家亚力山德拉儿童医院 Royal Alexandra Children's Hospital

Location: Brighton, UK **Completion date:** 2007 **Designer:** BDP **Photographer:** David Barbour, Sanna Fisher-Payne **Area:** 15,500 m^2

项目地点：英国，布莱顿 完成时间：2007年 设计师：BDP建筑事务所 摄影师：大卫·巴伯尔、桑娜·费舍尔·佩恩 面积：15,500平方米

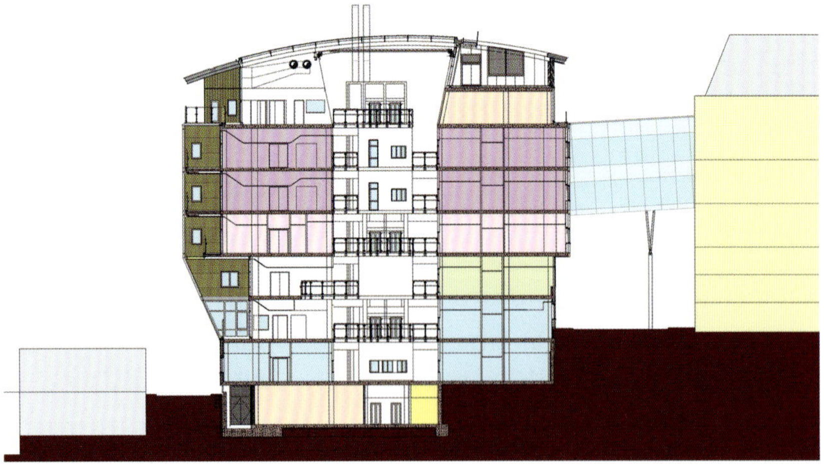

The site was a very tight handkerchief of land wedged between two roads and a number of existing buildings. Planning constraints dictated a maximum height. The very steep hill overlooking the Channel was the location's great saving grace, affording a sunny panoramic sweep over the sea and historic Brighton. The northern views from the site were towards an existing multi-storey car park and renal building, so the designers decide to give every bedroom a sea view and ensure that all inhabited rooms would have a window, including spaces like operating theatres which are traditionally windowless.

The design concept is "Children's Ark", an image that crystallised a number of important themes: from the idea of a "sustainable" community centred around the family to the nautical spirit of Brighton and the boat-like form of the building. With its soft, rounded corners and sheltering roof it projected a reassuring and optimistic image, around which the designers could integrate all the elements of the brief in a coherent and appealing way that would resonate with children, families and staff. At the centre of the hospital, the atrium binds all levels together, ensures good day-lighting throughout the building and creates a strong sense of the whole. The designers envisaged this heart space with the outpatients' reception and waiting area and adjacent café and shop as the social hub of the hospital.

The top decks of the "Ark" are the family and children's domain with the parents' accommodation and Play Centre opening out onto "play-decks" that enjoy the sea and sun. The staff lounge and doctors' mess are also located at the top, so that staff can have a break from the clinical environment and "take some air". The designers have provided balconies for play and relaxation at every level of the hospital, so that it makes the most of the wonderful prospect and gives primacy and visual expression to its users. The uniqueness of its identity is expressed externally through its form and the incorporation of bands of colour as well as by the irregular punctuation of child-scaled windows. This "individuality" is balanced by other attributes that blend the building into its context: the white pre-cast concrete panels that form the hull of the ark, echoing the prevailing white stucco of the city, anchored against a base of buff brick that ties in with some of the surrounding hospital buildings.

项目的场地十分狭小，位于两条路和许多建筑的夹缝之中，因此只能将建筑高度最大化。俯瞰英吉利海峡的陡峭山坡是项目所在地的唯一可取之处，可以俯瞰大海和布莱顿城的全景。项目的北面是多层停车场，因此设计师决定让每间房都能够看到海景，并且保证每间房都有窗户，包括传统意义上没有窗户的手术室。

设计理念围绕“儿童方舟”展开，这其中包含一系列的主题，从以家庭为中心的“可持续”社区到布莱顿的航海传统和建筑的船形造型。建筑柔和、圆润的边角和避风港似的屋顶营造出让人安心而积极的形象。设计师将项目所需的元素以一种连贯而吸引人的方式融入其中，与儿童、家属和工作人员达成了共鸣。医院的中庭将各个楼层捆绑在一起，保证了充足的采光，也营造了强烈的整体感。设计师将中庭、门诊接待处、等候区、咖啡厅和商店作为医院的社交中心。

“方舟”的顶层甲板上是家长住宿处和游戏中心，开放的“游戏甲板”是享受阳光和大海的好去处。员工休息室和医生食堂也位于顶层，工作人员可以从医疗环境中走出来，透透气。设计师在医院的各个楼层都提供了游戏和休息的阳台，具有良好的私密性和视觉感受。阳台的独特特征将在外形、色彩和儿童型窗户中得以体现。其他属性平衡了这种个性，将建筑与环境融为一体，例如：白色预制混凝土板形成了方舟的船体，与城市的白灰泥遥相呼应，浅黄色的砖底座与周边的历史建筑相得益彰。

1. You can enjoy the beautiful landscape through the windows inside the hospital
2. The surroundings
3. The soft and rounded architecture form

1. 从医院可以看到外部优美的景致
2. 医院周围建筑群
3. 柔和圆润的建筑造型

1

2

3

1

2

3

1. The rich and lively interior design elements
2. The atrium connects the different levels
3. Playful image on the wall

1. 丰富活泼的室内设计元素
2. 医院的中庭连接了各个楼层
3. 充满童趣的墙画

1. Main entrance
2. Fundraisers
3. Children's community nursing team
4. Physiotherapy
5. Outpatients department
6. Café

1. 主入口
2. 集资办公室
3. 儿童社区护理
4. 理疗区
5. 门诊部
6. 咖啡厅

1

2

1. The interior lighting is good
2. Medical area

1. 内部空间采光良好
2. 医疗区

康复中心儿科和工作治疗科 Children's Department and Work Therapy at the Institute for Rehabilitation

Location: Ljubljana, Slovenia **Completion date:** 2008 **Designer:** Dans Architects **Photographer:** Miran Kambic **Area:** 4,569 m^2

项目地点：斯洛文尼亚，卢布尔雅那 完成时间：2008年 设计师：丹斯建筑事务所 摄影师：米兰·卡姆比克 面积：4,569平方米

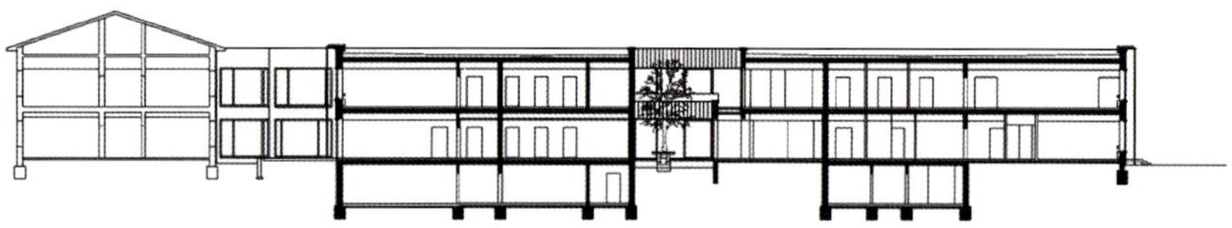

The whole institute complex was designed between 1954 and 1962 by who envisioned low and elongated buildings, characteristic of Scandinavia, a pavilion set in a green environment. The new Children's Department and Work Therapy building with the children's unit on the ground floor and the vocational rehabilitation unit on the first floor keeps in with the pavilion-like design. It is linked with the existing complex by means of the interjacent middle part made of concrete and glass housing the lifts. The building itself has a rectangular floor plan with an internal atrium. The corridors and halls are well-lit and interesting. Along the sides of the atrium and the corridors, there are living spaces that open outwards into the exterior space with their wooden terraces. The large windows of the rooms and other spaces overlook the old park with the large trees.

The simple architectural idea of a box in the middle of a park with a wooden atrium is based on the rational and functional plan. But this idea does not convey the whole identity of the architecture. The building tries to surpass the rigid functionality by using some fresh architectural ideas, for instance the wooden atrium, the playful graphics (by Slovenian artist Natasa Skusek) and the character of the exterior.

The building is built around the wooden atrium that enables the light to penetrate into the interior space. The wooden finish gives it a warmth and velvety softness. The users can use the atrium in a way it best serves their needs. The interior of the building is also graphically and sensorially exciting. The bold colours in the hallways with soft flooring serve as orientation around the building. The grey concrete walls of the core contrast the softness of the wood in the atrium, and the hygienic whiteness of the bathrooms is contrasted with bright coloured ceilings.

The building fits organically into the planted environment. The basic façade, made of fibre cement panels with a smooth finish, is white with large horizontal window apertures and a rhythm of vertical yellow accents made of corrugated panels. The volume of the building is perceived as a white canvas for projection of the graphics: the rhythmically placed yellow stripes, which should have been covered with an outer-green layer. The green layer would integrate the building into its environments and at the same time emphasise its poetics.

康复中心建于1954年和1962年之间，采用了经典的斯堪的纳维亚建筑模式，低矮而细长，处在绿树环抱之中。新建的儿科和工作治疗科楼保持了凉亭式结构，儿科位于一楼，职业康复科位于二楼。它通过中间带有电梯的混凝土玻璃连接处与原有的综合楼相连。建筑自称呈矩形布局，中间是中庭。走廊和大厅明亮而具有趣味性。中庭和走廊的两侧是活动空间，通过木板平台与外部空间相连。巨大的窗户可以远眺公园和高大的树木。

位于公园中的简单盒结构建筑拥有一个木中庭，具有合理而兼具功能性的规划。但是建筑并不局限于这一特点，试图通过新鲜的建筑理念超越死板的功能性，比如木中庭、有趣的图案（由斯洛文尼亚艺术家娜塔萨·思古塞克制作）和独特的外形。

建筑围绕着木中庭展开，让阳光渗透到室内空间里。木装饰提供了温暖而柔和的氛围。使用者可以根据自身的需求使用中庭。建筑的室内设计生动而令人愉悦。地面柔软的走廊里，大胆的色彩起到了指向作用。灰色的混凝土墙壁与柔和的木材形成了强烈的对比，而洗手间干净的白色与天花板上亮丽的色彩也相互对照。

建筑与植物茂密的环境有机地结合在一起。基本外墙由纤维水泥板组成，表面光滑，上面装有巨大的水平窗户和有节奏感的垂直黄色波纹板。建筑宛如一张白色的画布，上面绘着有节奏的黄色条纹，外层又装饰了绿色框架。绿化带让建筑与环境结合起来，同时也增添了诗意。

1. Wooden atrium
2. The interior design
3. The façade uses white and yellow as its main colours
4. There are corrugated panels with yellow stripes on the façade

1. 木制中庭
2. 室内设计
3. 建筑外立面采用了白色和黄色两种颜色
4. 外墙附有垂直黄色波纹板

3

4

1

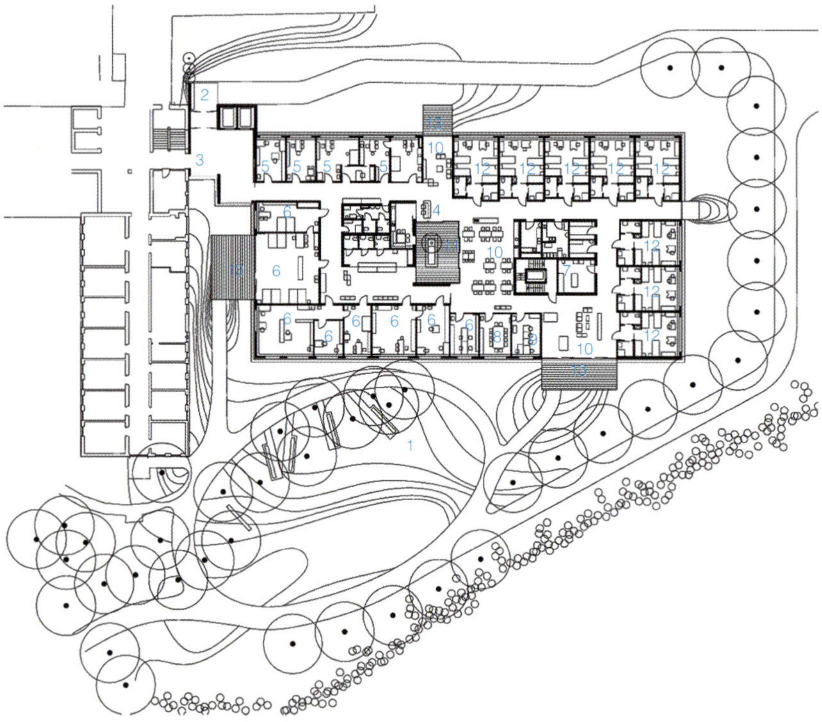

1. Park
2. Entrance
3. Accesss to the old building
4. Reception area
5. Doctor's office
6. Therapy room
7. Sanitary, cloakrooms
8. Seminar room
9. Classroom
10. Play and dining area
11. Inner atrium
12. Patient's room
13. Wooden terrace

1. 公园
2. 入口
3. 旧楼入口
4. 接待处
5. 医生办公室
6. 治疗室
7. 洗手间，盥洗室
8. 研讨室
9. 教室
10.游戏和就餐区
11. 室内中庭
12. 病房
13. 木平台

2

1. The use of wood brings security and intimacy for the patients
2. Bright colours enliven the interior atmosphere
3. Bold colours can be orientation tools

1. 木制材料的使用给病人带来安全感和亲切感
2. 彩色的使用增添了室内的活跃氛围
3. 大胆的色彩起到了指向作用

3

1. Grey concrete walls and colourful floors make a great contrast
2. The interior design is pleasant and lively
3. The pathway

1. 灰色的混凝土墙壁和彩色的地面形成鲜明的对比
2. 室内设计生动而令人愉悦
3. 通道

2

3

1

2

1. The rich and various design elements
2. Details of the patterns on the windows
3. The patient room is clean and simple
4. Public dining area

1. 设计元素丰富多样
2. 玻璃上的花纹细节
3. 病房布置简洁大方
4. 公共就餐区

3

4

Index 索引

irisarri+piñera
Doctor Cadaval 6-6I, 36202 Vigo, Spain
Telephone/Fax: 34 986 223875
Muros Health Centre

Kidosaki Architects Studio
104-0061 Ginza, Chuo-ku, 1-5-12, Tokyo, Japan
Telephone: 03-3562-2235
Fax: 03-3562-2237
Takekawa Hospital

Manuel Ventura & Associados Arquitectos
Rua de Santos Pousada, 350 4º Piso, 4000 – 478 Porto, Portugal
Telephone: +351 22 208 62 53
Fax: 351 22 208 93 15
Hospital CUF

Perkins Eastman
115 Fifth Avenue, New York, NY 10003, USA
Telephone: 212 353 7200
Fax: 212 353 7676
St. Vincent's Medical Centre
The Pavilion at Queens Hospital Centre

Perkins + Will
215 Park Avenue South, 4th Floor, New York, NY 10003, USA
Telephone: +1 (212) 251 7000
Fax: +1 (212) 251 7111
St. Mary's Hospital & Regional Medical Centre

Pitágoras Arquitectos
Rua João Oliveira Salgado 5 C - Costa, 4815-015 Guimarães, Portugal
Telephone: (+351) 253 419 523 / 253 519 149
Fax: (+351) 253 518 749
Guimarães Private Hospital

Richard Murphy Architects
15 Old Fishmarket Close, Edinburgh, Scotland, EH1 1RW, United Kingdom
Telephone: +44 (0)131 220 6125
Fax: +44 (0)131 220 6781
Stratheden 18-Bed Dementia and Mental Health Unit

RISCO architects
Av. do Brasil, 56 - 3º, 1700 - 073 Lisboa, Portugal
Telephone: +351 213 610 420
Fax: +351 213 610 422
Luz Hospital

RTKL
2101 L Street, NW, Suite 200, Washington, DC 20037, USA
Telephone: 202 833 4400
Fax: 202 887 5168
Community Hospital North Expansion

Samoo Architects & Engineers
890-6 Teheran Street, Seoul, Korea
Telephone: (02) 2184-5114
Fax: (02) 2184-5901
Seoul St. Mary's Hospital

Scott Wilson, in association with Capita
Scott House, Alencon Link, Basingstoke, RG21 7PP, UK
Telephone: +44 (0)1256 310 200
Fax: +44 (0)1256 310 201
Downe Hospital

Stanley Beaman & Sears
180 Peachtree St. NW Suite 600, Atlanta GA 30303, USA
Telephone: 404 524 2200
Fax: 404 524 8610
Kennedy Krieger Institute Outpatient Centre
Blanchard Hall – Roosevelt Warm Springs Institute for Rehabilitation

TEAM M Architekten ZT GesmbH
Eisenhandstraße 13-15, 4020 Linz, Austria
Telephone: +43 / (0) 732 / 78 43 81
Fax: +43 / (0) 732 / 78 43 81 - 24
Provincial Hospital Steyr

Tönies+Schroeter+Jansen Frei Architekten GmbH
Curtiusstraße 19, 23568 Lübeck, Gemany
Telephone: 0049 451 799 73-0
Fax: 0049 451 755 41
Johannes Gutenberg University Mainz, Conservative Medicine

UArchitects
Klokgebouw 233 5617 AC Eindhoven, the Netherlands
Telephone: +31 40 2366535
Marishof in Maarheeze

Ward+Blake Architects
200 E. Broadway, PO Box 10399, Jackson, WY 83002, USA
Telephone: 307 733 6867
Teton Radiology – Madison Imaging Centre

Zeidler Partnership Architects
315 Queen Street West, M5V 2X2, Toronto
Telephone: 416 596 8300
Fax: 416 596 1408
Assuta Medical Centre

ZGF Architects LLP
Portland Office, 1223 SW Washington Street, Suite 200,Portland, Oregon 97205, USA
Telephone: 503 224 3860
Fax: 503 224 2482
Providence Cancer Centre
Denver Children's Hospital

图书在版编目（CIP）数据

医院建筑 / 于萍编 ; 常文心译. -- 沈阳 : 辽宁科学技术出版社, 2011.10
ISBN 978-7-5381-7107-5

Ⅰ. ①医… Ⅱ. ①于… ②常… Ⅲ. ①医院－建筑设计－英、汉 Ⅳ. ①TU246.1

中国版本图书馆CIP数据核字(2011)第170371号

出版发行：辽宁科学技术出版社
（地址：沈阳市和平区十一纬路29号　邮编：110003）
印 刷 者：利丰雅高印刷（深圳）有限公司
经 销 者：各地新华书店
幅面尺寸：245mm × 290mm
印　　张：36
插　　页：4
字　　数：80千字
印　　数：1～2000
出版时间：2011年 10 月第 1 版
印刷时间：2011年 10 月第 1 次印刷
责任编辑：陈慈良
封面设计：赵　聪
版式设计：赵　聪
责任校对：周　文
书　　号：ISBN 978-7-5381-7107-5
定　　价：268.00元

联系电话：024-23284360
邮购热线：024-23284502
E-mail: lnkjc@126.com
http://www.lnkj.com.cn
本书网址：www.lnkj.cn/uri.sh/7107